RETHINKING COLD WAR CULTURE

Peter J. Kuznick

James Gilbert

RETHINKING COLD WAR CULTURE

Edited by Peter J. Kuznick
and James Gilbert

Smithsonian Institution Press
Washington and London

Copy and production editor: Ruth W. Spiegel
Designer: Amber Frid-Jimenez
Indexer: Andrew L. Christenson

Library of Congress Cataloging-in-Publication Data
Rethinking Cold War culture / edited by Peter J. Kuznick and
James Gilbert
p.cm.
Includes bibliographical references and index.
ISBN 1-56098-869-X (alk. paper)—ISBN 1-56098-895-9 (pbk: alk. paper)
1. United States—Civilization—1945. 2. Cold War—Social aspects—United States. 3. United States—Social conditions—1945. 4. United States—Intellectual life—20th century. 5. Popular culture—United States—History—20th century. 6. Poltical culture—United States—History—20th century. I. Kuznick, Peter J. II. Gilbert, James Burkhart.
E169.12.R475 2001
973.91—dc21 00-045015
British Library Cataloguing-in-Publication Data available

Manufactured in the United States of America
08 07 06 05 04 03 02 01 5 4 3 2 1

∞ The recycled paper used in this publication meets the minimum requirements of the American National Standard for Information Sciences—Permanence of Paper for Printed Library Materials ANSI Z39.48-1984.

Contents

Acknowledgments

This volume originated with the Landmarks Conference on Cold War Culture sponsored by the History Department at American University and the Smithsonian Institution's National Museum of American History, at which early versions of many of these chapters were presented. We would like to thank both institutions for their generous support and particularly acknowledge the special contributions of Gary Kulik, Art Molella, Betty Bennett, Roger Brown, and Jennifer Winter. We are also grateful for the insightful comments of Lily Kay and Nelson Lichtenstein. Finally, we would like to thank Mark Hirsch and Ruth Spiegel of the Smithsonian Institution Press for their encouragement and assistance.

U.S. Culture and the Cold War

Peter J. Kuznick and James Gilbert

In the 1954 film *Strategic Air Command,* Dutch (Jimmy Stewart), a star pitcher for the St. Louis Cardinals, tells his wife, Sally (Doris Day), that he has decided to forsake his lucrative baseball career and stay in the Air Force. Sally, not understanding, protests this decision on the grounds that there is no war going on. Dutch corrects her: "But there is a kind of war. We've got to stay ready to fight without fighting. That's even tougher. That's why I made this decision." Sally blurts out: "You made it. We didn't."[1]

But Dutch was right, as Sally eventually came to appreciate. The United States was engaged in a new kind of war—a war that required constant vigilance and readiness to fight on a moment's notice. Labeled the "Cold War" in 1946 by Bernard Baruch, this war was unlike any other the United States had ever fought. For one thing, it lasted a lot longer than any previous war. Historians debate the date of its inception but, for our purposes, August 6, 1945, the day the United States dropped the first atomic bomb on Japan and introduced nuclear terror to the world, is as good a starting point as any. Though many features of Cold War culture were given their most heightened expression in the 1950s, the developments we are discussing were in no way confined to that decade or the peculiar patterns of thought and behavior commonly associated with it. While the Cold War lasted until the fall of the Soviet Union in 1991, much of what is usually thought of as Cold War culture outlasted the Cold War itself and likely will be with us for a long time.

Although the Cold War shaped and distorted virtually every aspect of American life, broadly speaking there is very little fundamentally new about American culture in the Cold War era. Most of the characteristics by which we define it are the results of long-term social trends and political habits of mind, revived and refurbished from the past.

Yet interacting with long-term trends, which we will discuss, four new elements, unique to the Cold War era, dramatically transformed aspects of American politics and culture. These new elements—threat of nuclear annihilation, replacement of direct military confrontation with surrogate and covert warfare, opposition to an enemy officially espousing socialism and supporting Third World revolution, and the rise of the military-industrial complex—were very closely related though not identical.

More than anything else, nuclear weapons changed the world, introducing an element of vulnerability that had not previously existed. Although Harry Truman responded to the atomic obliteration of Hiroshima by exulting, "This is the greatest thing in history,"[2] many Americans were not so sure. In first reporting the bomb, popular radio news commentator H. V. Kaltenborn sounded a much more cautionary note, warning, "For all we know we have created a Frankenstein! We must assume that with the passage of only a little time, an improved form of the new weapon we use today can be turned against us."[3] When the Japanese surrendered the following week, Edward R. Murrow observed that "seldom, if ever, has a war ended leaving the victors with such a sense of uncertainty and fear, with such a realization that the future is obscure and that survival is not assured."[4]

The intensity of nuclear fear ebbed and flowed during the course of the Cold War, but the threat of annihilation was never far from the public mind. Broad-based concern burst forth in the immediate postwar years as the campaign for international control of nuclear weapons floundered. The nuclear stakes escalated over the next decade, marked by the first Soviet atomic bomb test in August 1949, the United States and Soviet hydrogen bomb tests in 1952 and 1953, the heightened mid-1950s concern about nuclear radiation and fallout, and the apocryphal missile gap warnings that followed the Soviet launching of Sputnik in October 1957. Nuclear terror peaked in October 1962 when the United States and the Soviet Union came within a hair's breadth of war during the Cuban missile crisis.

Nuclear fear abated somewhat between 1963 and 1980, a period historian Paul Boyer has labeled the "Era of the Big Sleep,"[5] but was rekindled dramatically by the 1980s missile-rattling of Ronald Reagan. Reagan's bellicose rhetoric and Star Wars proposals aroused dormant nuclear anxieties. Hollywood responded with a spate of nuclear war movies. The 1983 ABC television production *The Day Af-*

ter, the second most-watched television event ever, sparked an unprecedented national debate. Scientists warned that even limited nuclear war could trigger nuclear winter and end all life on the planet, a view echoed in widely read books like Jonathan Schell's powerful best-seller *The Fate of the Earth.*

This possibility of a nuclear holocaust that would spare no one—not even its perpetrators—provided impetus for a second new element in this period: surrogate and covert warfare. Over forty-five years the United States fought two bloody surrogate wars against the Soviet Union and China, one to a draw and one ending in defeat. In the Korean War (1950–1953), U.S. forces, under cover of the United Nations, battled against the North Koreans and the communist Chinese. The war legitimized a tremendous increase in defense spending and the further militarization of science and technology. The war also reinforced American willingness to intervene internationally against governments and political movements, both democratic and undemocratic, that were deemed a threat to American interests. While one could trace the United States becoming an antirevolutionary force on the global stage to the 1890s, U.S. opposition to radical movements in the Third World reached unprecedented heights during the Cold War era. Beginning with U.S.–backed overthrows of popular governments in Iran in 1953 and Guatemala in 1954, the United States and the Soviet Union became involved in dozens of covert and not-so-covert wars over the next four decades, with both sides providing weapons, advisors, and financing.

In 1954, on the heels of the stalemate in Korea, the United States replaced the defeated French colonial forces in Indochina. American troop commitment escalated rapidly in the mid-1960s. By the time the Americans were forced to evacuate in 1973, a total of 2.7 million American men and women had been sent to Vietnam. Fifty-eight thousand Americans and nearly four million Vietnamese were killed. Although some Americans would cling to the fantastic notion that the United States fought with one hand tied behind its back, records show that the U.S. dropped more tons of bombs on tiny impoverished Vietnam than had been dropped by all sides combined in all of history's previous wars.

America's nuclear buildup, global mobilization, and interventionism during the Cold War were justified in the name of stopping Soviet communism, a foe policymakers deemed so diabolical that its defeat warranted the risk of destroying civilization itself. The fact that the Soviet Union was not just a military, economic, and geopolitical but an ideological foe, committed in theory, though clearly not in practice, to social and economic equality and the socialization of the means of production, posed a unique kind of challenge to a resolutely capitalist nation. This ideological antagonism between socialism and capitalism polarized the world along new lines. In the Third World this often created a situation in which the Soviet Union was supporting revolutionary forces against U.S.-armed and backed military dictatorships. In the United States political debate often became skewed and distorted, as conservatives easily discredited social democratic or left-wing solutions to societal problems by identifying their proponents with the brutal and repressive Soviet system—a system that most liberals and leftists themselves deplored. Saddled with the Soviet albatross around its neck, a once-vibrant American left was effectively marginalized, narrowing and transforming the terms of debate for much of the Cold War era and leading—except for a brief, though profound, moment in the 1960s—to the loss of idealism and the virtual death of Utopianism. It is thus remarkable how thoroughly the United States had repudiated the values underlying the liberal construction of World War II as the democratic war against racism, fascism, and colonialism. In fact, between the late 1940s and mid-1960s, those who still stubbornly clung to such "subversive" values risked being ostracized or more seriously persecuted.

To summarize, there were several cycles of military, ideological, and psychological warfare that sometimes overlapped but did not always converge. During the period from the late 1940s through 1963, all three of the forces propelling Cold War culture exerted influence. For the next decade, dominated by surrogate warfare in Vietnam, anticommunism remained high, but nuclear fear abated somewhat. After a brief respite during the Ford and early Carter years, the Cold War reheated between 1978 and 1988, as Carter, to a limited extent, and Reagan, to a great one, brought back the surrogate warfare, ideological crusades, and nuclear posturing that had marked the 1950s. Under Reagan, the possibility of nuclear warfare again shrouded the nation with nightmares of Armageddon. Presidents Bush and Clinton called for new approaches, but did little to reverse the habits of mind that the Cold War created. So defense spending remains at near Cold War levels and strategy continues to rely on nuclear weapons.

All this notwithstanding, the long age was by no means one of unremitting gloom and despair, far from it. The four new factors—surrogate and covert warfare, nuclear fear, anti-Sovietism, and a powerful military-industrial complex—interacted with other forces that some observers have mistakenly attributed to the Cold War. Like all other eras of American history, this too shares a complex order of long- and short-term trends, immediate and distant causes, cycles of intensity, and inconstant speed of changes.

In this chapter as in the book, we hope to make sense of three major developments: (1) The varieties of long-term historical trends that come together in the postwar world to shape its culture. (2) The ways in which change was cyclical or at least uneven, speeding up at one point and slowing at another. (3) The manner in which the four unique elements of the Cold War impinged upon and affected long- and short-term historical changes. But what were the long-term trends with which these forces interacted?

To begin with, there is demographics. The most intimate and perhaps meaningful decisions of American adults—to form families and have children—seem remarkably untouched by the Cold War. The amazing rise in the birth rate during the late 1940s and 1950s (peaking in 1957), the declining age of marriage and growing marriage rate, and the low divorce rate all converged in the midst of the most intense years of nuclear fear and ideological and surrogate warfare. While it might be possible to interpret this reaction as hunkering down in the face of doom—retreating within the four walls of the household—it is far more likely that these were optimistic measures in response to the despair of the 1930s and the enforced separation of young Americans during the war years. It might even be argued that the 1950s was the first decade in which the nuclear family finally emerged universally in America, after decades of extended family living arrangements.

Since family formation and fertility respond to positive and negative economic and cultural stimuli, it is not surprising that this era of comparatively good times brought increased marriage and fertility. Furthermore, increased sexual activity at younger and younger ages, especially for women, first manifested itself in very early marriage and, then, after the 1950s, in growing incidence of premarital sex. By the 1990s this meant a much lower marriage rate, increased age at marriage, a declining birth rate, and the initiation of an active, adult, sexual life at a much earlier age. While the heightened uncertainty about a future threatened by nuclear war reinforced the tendency to live for the moment rather than defer sexual and other gratifications, the sexual revolution would probably have occurred without the Cold War.

The history of sexuality in postwar America is thus the story of increasing liberalization, first within and then outside of marriage. This occurred in three phases: the rebellion of men against marriage and the increasingly permissive fantasy life associated with the "Playboy" lifestyle of the late 1950s, the more significant rebellion of women against prescribed sexual expectations and behavior, and the increasingly open display and public tolerance of homosexuality beginning in the late 1960s. Exploited by market forces, particularly advertising, facilitated by improved contraception (especially the introduction of the birth control pill in 1960), and promoted by Hollywood films and popular music, this heightened and seemingly ubiquitous sexuality produced the related challenge to the traditional nuclear family. By the 1980s, a backlash was created among conservative forces which, through banning abortion and promoting "family values," sought to reverse this trend.

If all of these changes conspired to raise the average age of marriage and lower the age of sexual initiation over the whole period, another result has been to lengthen greatly the stage of adolescence in America and to transform it from a period of social, economic, psychological, and cultural dependence upon the family into a period of dependent adulthood. Again the 1950s played a key transitional role in this transformation of the precocious adolescent from a transgressor against adult privilege and behavior to a credit-card-carrying member of a loose family organization. If anything, the gradual lowering of the age of sexual activity and liberation of sexual mores unify this longer period far more than the fluctuations in marriage rates and ages divide it.

Just as family-related demographics and sexual behavior changed rapidly after 1945, so too did the physical movement of Americans out of older eastern and midwestern cities into the suburbs. Huge population shifts occurred as Americans moved from these same, older urban areas into the West, the Southwest, and the South (particularly Florida, Georgia, and the Carolinas). While this movement may have looked new because it was so dramatic, it was part of an older trend that had its roots in the population movements of the 1930s and particularly in the military dispersal of millions of young Americans during the Second World War. A nation is never so mobile as when it is at total war fighting enemies from both coasts. So the 1950s shift in the Cold War economy sped up a process that had begun earlier and continued well after the social and cultural observers wrote their amazed accounts of its implications. Thus when Lucy and Desi Arnez and Fred and Ethel Mertz abandoned their (television) apartments in New York City, they were only following what millions of ex-G.I.s and their wives had been doing since World War II, and what the film industry had done four decades earlier when it left studios in New York and New Jersey for sunnier climates.

Looked at over the whole of the twentieth century, the communications revolution of the 1950s endlessly portrayed and fictionalized contemporary culture. It brought television into millions of American homes but in fact was only one of a series of dramatic changes in communications marking almost every decade of the century. In the twentieth century, there have been at least three movie revolutions: the pre–World War I era of flickering nickelodeons, the ambitious and glamorous silent pictures of the 1920s, and the revolution in sound and color of the 1930s. Throughout these decades radio, too, dramatically changed in America. Then came the impact of television in two phases—an introductory half-decade or so after World War II, and a second, intense period of change during the 1950s and 1960s when television became universal and its full potential became apparent for the first time. Add to this the deepening sophistication in recorded sound and the dimensions of the communications revolution become clear. With plummeting prices, television, radio, records, and music became available to almost every American. Yet this earlier communications revolution pales beside a second intense age of invention and distribution of culture based upon computers, digitalized sound and images, satellites, and other advanced technologies that began in the 1980s. Next to this, the 1950s television revolution seems to be primitive at best, and remarkable only because the changes were so visible and sparked so much discussion about its effects on American society.

Democrats and Republicans have dominated politics throughout the postwar era, successfully fending off or coopting most challenges from the left and the right. Third parties have found it almost impossible to crack the two-party system. Within the two major parties, a general consensus prevailed on both domestic and foreign policies. Truman's ouster of former Vice President Henry Wallace as Secretary of Commerce in September 1946 not only completed his purge of New Dealers from his cabinet but effectively ended the vigorous postwar challenge from his own party's once-formidable left wing. Thereafter political debate and electioneering proceeded within narrow ideological bounds, especially where foreign policy was concerned. The basic framework of international containment and limited social democracy—guns and butter—held sway until challenged by the antiwar movement in the late 1960s and by Reagan-Bush efforts to dismantle the welfare state in the 1980s.

Overall there has been a steady decline of the Democratic majority and coalition assembled in the late 1920s and 1930s. The Republicans have gradually and very slowly become the majority party in the United States—if not by actual affiliation, then by voting behavior, especially at the presidential and congressional levels. This has been, for much of the era, a very bitter struggle, occasionally over serious philosophic and policy issues but more often between men of great political ambition and lesser principles.

What is not new or unique to politics of the Age of Cold War is character assassination, opportunism, anticommunism, smear, and guilt by association—in other words, "McCarthyism"—although its late 1940s and 1950s practitioners, including not only Nixon Republicans but Harry Truman and many of his Democratic colleagues, took these practices and turned them into an art form. During its most virulent phase, the firing of respected professors, including top scientists from major universities, blacklisting of Hollywood celebrities, persecution of policymakers and lower-level government employees, targeting of homosexuals, and celebration of the stoolpigeon as the citizen par excellence, induced widespread fear and temporarily silenced some of the nation's most creative voices. American democratic politics has periodically bottomed out on the lowest level of the standard "all that the traffic will bear." Free enterprise and market principles translated into politics has meant a boom and bust cycle of ethics.

Anticommunism was first effectively employed against labor and radical forces in the aftermath of the Haymarket episode in 1886 and again on a large scale during and after World War I. Anticommunism turned against Democrats was a well-known tactic of the 1930s. The Democrats circulated their own kind of debased coin in identifying every Republican, beginning with Herbert Hoover, with the Great Depression. What is remarkable is how little anticommunism really worked to reverse the Democratic political majority after World War II. That process was much more gradual and depended upon complex long-term phenomena.

One of these was the state of the economy. In the short run, the economy always has an immediate impact upon elections and other forms of political culture. In the long run, the economy's impact on American life was even more pronounced. The period following World War II, which began by unleashing the enormous pent-up demand for housing, transportation, and other consumer goods and services such as higher education, is characterized by generally strong growth and the development of a modern consumer economy. Within this broader age, the 1950s was the most egalitarian decade of the postwar period as measured by distribution of wealth. Since then inequality has steadily increased, leading to the tremendous disparity between rich and poor of the Reagan-Bush-Clinton era. During this period, too, there has been a remarkable transformation of the American economy. It could be argued that the 1950s is the last decade of the industrial age. Thereafter the American economy shifted rapidly to producing advanced technology, services, distribution and credit systems, entertainment and communications, banking, and construction. Basic manufacturing industries, on the other hand, declined and relocated in search of lower-paid and nonunion workers.

Population distribution reflected this transformation. From the four-part urban model—downtown commercial, retail, and entertainment center; middle-class residential area; industrial zone; and surrounding working-class housing—the line between cities and suburbs disappeared. New sorts of population conglomerates and nodules merged with multiple smaller centers. Dispersed populations were linked by highways, and engaged in many new small-scale white-collar forms of work and commerce. Among the most striking developments in the 1950s is this rapid transition to a new pattern of work, accommodation, and consumption.

Within this transformation of so many aspects of American life, the most profound social alteration was probably the end of the legal segregation of the races in the South and the tentative beginning of integration throughout the nation. A social goal still unachieved—the color-blind society—has had many beginnings in America and many moments to commemorate. Throughout the century and a half since the Civil War, high expectations of progress have repeatedly been dashed and betrayed by political opportunism and racism. Yet a variety of forces came together at the end of World War II that doomed legal segregation forever in the South and held out hope for fundamental restructuring in the North. After making significant strides in the 1950s, civil rights activists anticipated even greater gains in the 1960s. The walls of legal separation, so carefully constructed since the end of the nineteenth century, crumbled rapidly. Efforts to eliminate de facto discrimination and resulting social and economic inequality met resistance, however, from hostile whites who were goaded in part by uncertain employment prospects and a steady two-decades-long decline in real wages beginning in 1973. Since then, each decade of the postwar era has been faced with some new sort of crisis related to the painful process of eliminating prejudice and inequality in all their tenacious forms.

Something of the same story might be told about the position of women in America during these years. Despite the weight of advice, opinion, and propaganda, women continued (after a short recess immediately after 1945) to enter the work force in larger and larger numbers. Women's changing socioeconomic position burst into public view in the early 1960s when feminists denounced the suburban housewife lifestyle that kept women dependent, undereducated, underemployed, and largely underfulfilled. Although some deliberately exaggerated the prevalence and harmful effects of a lifestyle that had never been universal nor, by and large, entirely imposed on those who lived it, their critique had a salutary effect. Not only did it create a past to be universally despised and rejected; it proposed new ways of organizing gender relations that had a profound effect on the very real inequalities that existed everywhere between men and women. For the next three and a half decades, perhaps no issue would prove so divisive—and creative—as this ongoing effort by growing numbers of women to change their social roles, redefine the nuclear family, and assert increased control, real and symbolic, over their bodies.

Advocates of women's liberation often encountered stiff resistance from traditionalists whose social conservatism was reinforced by the religious awakening that began in the United States during the 1950s and, with some ups and downs, continues to the present. This longest awakening in American history—a history that is awash with eruptions of religious fervor and evangelical agitation—is also one of the most remarkable. Catalyzed by the new methods of communication developed by Bishop Fulton J. Sheen, Norman Vincent Peale, and Billy Graham, this impulse has moved significantly from the periphery of American politics to its very center.

Religion has always played a large and sometimes contradictory and complex role in American politics. Throughout most of the twentieth century, for example, the Protestant majority worked to exclude Catholics and Jews from high national office. Yet on a local level, Catholic and Jewish groups could often control ethnic-religious urban political machines. Religion also played a significant role in the Populist movement of the late nineteenth and early twentieth centuries and again in the 1920s. In the postwar world, religion emerged gradually and steadily as a potent political force. To some degree this was even true of the presidency of (Catholic) John F. Kennedy. Even more important was the emergence of second-wave, second-generation Protestant evangelicals (after Billy Graham) who merged conservative theology with conservative politics behind such rallying cultural issues as abortion, but cut much deeper into the cloth of the social welfare state by proposing a sort of neoliberalism of the nineteenth-century variety.

America's welfare state had begun during the New Deal and expanded rapidly in the immediate postwar period into large-scale government intervention into society, culture, and economics. The G.I. Bill of 1946, for example, was a landmark extension of the principles of the welfare state as was the expansion of such programs as social security. Very similar was the introduction of a host of other new programs that were passed up to the end of Richard Nixon's term in office. Thereafter from 1973, despite some additional gains, optimism about the possibilities of a socially engineered society eroded. Declining real wages and dwindling opportunities turned erstwhile allies into fierce competitors over shrinking benefits.

If these cycles and long-term developments define the postwar period in a broad sense, what is the specific role of the unique factors of the Cold War? How did nuclear fear, anti-Sovietism, surrogate and covert wars, and the military-industrial complex alter the transitions and changes that flow inexorably through this period? Clearly their impact was substantial, particularly in two clusters of years: from the end of the 1940s to the early 1960s, and in the first half of the 1980s. But Cold War culture is not synonymous with American culture, even at the height of its impact. It is the interaction between these unique elements of Cold War culture and the long-standing trends that existed independently and in large part antedated the Cold War that created American civilization in this age.

We take strong issue with those observers who have found the Cold War to be responsible for every change and cultural distortion occurring during these years. Nevertheless, the vividness of the perceptions suggests that the principal effect of the Cold War may have been psychological. It persuaded millions of Americans to interpret their world in terms of insidious enemies at home and abroad who threatened them with nuclear and other forms of annihilation. Seeing the world through this dark, distorting lens and setting global and domestic policies to counter these fanciful as well as real threats was and is, then, the largest impact of the Cold War.

The chapters in this book explore the ways in which the Cold War intersected ongoing developments of American civilization and sometimes added unique elements of its own. In doing so, they challenge and complicate many of the previous interpretations of this period. Readers will, we believe, quickly become aware of how difficult and contentious—and how exciting—this undertaking is.

As the chapters show, Americans experienced the Cold War in a multitude of ways that reflected profound differences between generations, genders, races, classes, regions, and religious groups. William Tuttle and Peter Filene explore the impact of generation. Tuttle finds that children raised during the war and in its aftermath were pulled in dramatically different directions. The Holocaust and the atomic bomb created an immediate sense of their vulnerability in a world whose end was imaginable. At the same time, permissive child-rearing created the expectation of being nurtured and protected, heightening the value of life. These contradictions begat a generation that both sought security and rebelled against it.

But generational difference isn't everything. Peter Filene, in analyzing the impact of the Cold War on those raised during the Great Depression and the Second World War and also on their children, highlights differences between members of the same-age groups. He shows that while postwar policymakers became obsessed with the threat of communism at home and abroad, most Americans were more concerned with work and family. In the 1960s their children, raised to expect security and fulfillment, rebelled against obstacles posed by sexual puritanism, rigid gender roles, and the United States invasion of Vietnam.

All of the postwar generations grew up in a new economic environment. Ann Markusen demonstrates that the needs of a defense-based economy spurred growth but distorted existing patterns of work and community. Establishing that the military might be needed to insure that America's form of industrial capitalist democracy prevailed in the Cold War precipitated the rise of gunbelt regions in the south Atlantic states, New England, and the mountain and Pacific states. In fact, she argues, not only did traditional industrial areas decline as a result, but almost no aspect of American life emerged untouched by the force of this investment and production. Even politics shifted as conservatives from defense-rich areas exercised greater and greater influence.

The resulting polarization of American politics shattered the illusory though deep-seated sense of unity that pervaded American culture during the 1950s. Alan Brinkley identifies the factors that gave rise to this illusion as postwar prosperity, which spawned a huge new middle class, suburbanization, increasing bureaucratization of white-collar work, and proliferation of middle-class images and values through television westerns, variety shows, quiz shows, and situation comedies. None of these elements, Brinkley contends, flowed directly from the Cold War itself. But the stiff uniformity of this new middle-class culture provoked a scathing critique from writers and intellectuals.

Other groups rebelled as well. Joanne Meyerowitz challenges the belief that the 1950s were uniformly conservative and bleak by identifying sources of dissent within Cold War culture itself. A number of groups, including women and gay men and lesbians, in order to claim expanded rights and privileges, circumvented the conservative logic of Cold War culture by appropriating the very language and values U.S. officials used to promote the Cold War. This tactic was also used effectively by African Americans who tried to force the United States to live up to the human rights promises of Cold War rhetoric.

Jane De Hart elaborates a different interaction between gender roles, sexuality, and anticommunism. Showing how American nationality and the human body have become integrated and parallel metaphors, she examines the special pressures the Cold War placed upon women. Domestic containment, or the promotion of an exaggerated domesticity designed to thwart a budding revolution in gender roles and expression of "deviant" sexuality, became an almost official ideology. And yet, De Hart argues, millions of women maneuvered successfully to escape its constraints. De Hart further demonstrates that while Cold War containment of international communism gave domestic containment its unique coloration, efforts to undermine the sexual and gender revolution have outlasted the Cold War and proved that the link was more convenient than inherent.

National identity during the Cold War was also grounded in American military superiority. But the novel *Catch 22* exposed the fundamental irrationality of an economy and culture based upon the insane logic of militarism, particularly in the nuclear age. Stephen Whitfield contends that Joseph Heller's epic novel was as much about the Cold War as it was about World War II where the action is set. It had its greatest resonance with a generation appalled by the seemingly inexorable expansion of the war in Vietnam. Heller's brilliance was to reverse the values and assumptions of warfare. Heroes desert. The rules of military conduct are corrupt. The purpose of war is irrational. The game is fixed.

Heller stood almost alone among writers in identifying the irrationality of Cold War logic in 1961. Among filmmakers, Stanley Kubrick occupies a similar place for his 1964 film *Dr. Strangelove,* which ridiculed the assumptions of deterrence and the nuclear arms race. But a much more common attitude toward the military in the 1950s was reflected in what Chris Appy labels "sentimental militarism." Appy discovers a moral certitude about American militarism, cloaked in a curtain of benign nostalgia, in several of the most popular movies of the decade. These films tried to create popular support for the ongoing militarization of American society and its struggles in places like Korea. But, Appy contends, this World War II–style sentimentalism even rang false at the time. Its cheery imbecility would soon be exposed by Heller and Kubrick and thoroughly discredited by opponents of United States overseas aggression.

When Leo Ribuffo steps back even further, viewing the Cold War era in the context of century-long trends, he finds surprisingly little that is new. While other contributors to this volume have located the roots of Cold War society, culture, and politics in the 1930s and 1940s, Ribuffo traces most of them to Progressive politics and Woodrow Wilson's foreign policy. He ends by challenging us to reconsider what if anything has actually changed.

Historians and other citizens will have to answer the vexing questions posed by Ribuffo and the other contributors to this volume. The understandings we derive will be of more than academic interest. In a crucial sense, how we understand this troubling though fascinating period in human history will go a long way toward determining the kind of future we collectively shape in the twenty-first century and beyond.

Notes

1. *Strategic Air Command,* director, Anthony Mann, producer, Samuel J. Briskin (Paramount, 1955).
2. Harry S. Truman, *Memoirs: 1945 Year of Decisions* (New York: Doubleday, 1955; Signet Books, 1965), 465.
3. Quoted in Paul Boyer, *By the Bomb's Early Light* (New York: Pantheon Books, 1985), 5.
4. Quoted in Boyer, *Bomb's Early Light,* 7.
5. Boyer, *Bomb's Early Light,* 355.

CHAPTER 1

America's Children in an Era of War, Hot and Cold

The Holocaust, the Bomb, and Child Rearing in the 1940s

William M. Tuttle Jr.

hen did the Cold War begin? This is an issue historians have debated for years. Did it begin in 1947, when President Harry Truman proclaimed the containment doctrine, or in 1948 when the Soviets overthrew the elected government in Czechoslovakia? Or did the Cold War begin earlier, perhaps when the United States dropped the atomic bomb in 1945 or, as some historians have suggested, still earlier because of deepening suspicions among the members of the Grand Alliance as victory approached and an uneasy postwar world loomed?

For America's children, the answer is easy. Both the Cold War crusade against an absolutely evil enemy and the nuclear specter—or at least the haunting fear of dying in an air raid—began during the Second World War. These phenomena, which are related, stemmed from the air raid drills and blackouts, the newsreels and war movies, the photographs of death in *Life* magazine, and in the countless war games that children played in vacant lots and on school playgrounds. Too often historians have overlooked the continuities that distinguished the decade of the 1940s, at least in children's lives. Between 1941 and 1945 the specter of death and destruction haunted millions of American girls and boys, and the way in which the war ended only exacerbated children's fears and insured that the specter would persist into peacetime. The war culminated in victory as exemplified by the boisterous celebrations on V-J Day on August 14, 1945, but it also ended in the stench of Dachau and Hiroshima.

It is well known that the emotions of war do not immediately dissipate with the armistice. Succeeding World War I, for example, was the Red Scare, which was an extension of the wartime atmosphere with its cult of patriotism and need for an enemy. The anti-German and anti-Bolshevik agitation of late 1918 and the early months of 1919 so overlapped that, as the historian John Higham has noted, "no date marks the end of one or the beginning of the other." By mid-1919, many Americans were no longer fearful of the German enemy but, emotionally, they were still at war. The homefront during the war had seen both mob action against German-Americans and government imprisonment of dissenters. After the war some Americans displaced their hatred onto a new enemy, an "inner enemy" composed of aliens and dissenters who should be—and soon were—deported by the federal government. To those Americans, the wartime and postwar enemies had become blurred: "Wonder if Berlin Bill is raising a scrubby beard so as to join the Bolsheviki?" asked *The Fleet Review* in July 1919. "He'd doubtless feel at home with his friends Trotsky and Lenin."[1]

So too it was in the aftermath of World War II. Within a year or two, Adolf Hitler's personification of the enemy had merged with that of Joseph Stalin; the result was "Red Fascism," which the historians Les K. Adler and Thomas G. Paterson have described as the assumption "that Russia and Stalin in the cold war would always act in a manner similar to Germany and Hitler" during the Second World War. Certainly President Truman held this viewpoint, as did most of the American people. "There isn't any difference in totalitarian states," Truman said in 1947. "I don't care what you call them, Nazi, Communist, or Fascist."[2]

To understand American children's lives in the years following the Second World War, it is necessary to look first at their lives during the war, particularly their attitudes and values as well as their popular culture and their private fears. Secondly, it is essential to understand the psychic havoc suffered by children as a result of their exposure to the enormity of two events in 1945: the Holocaust and the atomic bomb. America's children were forever changed by their awareness both of the death camps and the fact that, in seconds, a single bomb could incinerate an entire city—perhaps their hometown—and everybody in it. Finally, however, for many children living in postwar America there was a mitigating factor, one that provided much comfort in an uncertain world. During the 1940s there was a transformation in American child rearing. Although this change was underway in the late 1930s, the war's democratic ideology caused it to accelerate. "If one can judge a time from its books," Daniel Beekman, a historian of American child rearing, has written, "the 1940s during the war and in the years immediately afterwards may be described as the most permissive time in child care."[3]

Many observers at this time, as well as historians afterward, have written that one era had ended in 1945, on V-J Day, and another had begun. In many ways this was true. But for America's children, life went on; and important changes that began or accelerated during World War II persisted throughout the Cold War years to follow. This study examines these continuities in children's popular culture, in their private horrors, and, not least, in the institution of child rearing in the United States.

For America and its children the Second World War was a time of rapid social change. Uprooted by their families, millions of girls and boys had migrated to war-boom towns and army posts across the country. Regardless of location, the children had had to grapple with fears generated by the war. When the sirens sounded during the blackouts, homefront boys and girls worried whether this was another drill or, in fact, a real attack.

Children's fears began on December 7, 1941, when they saw that their own parents were afraid. Many girls and boys had never seen their fathers cry before, and some were frightened by the anger displayed by parents and other adults who raged against the Japanese. In grade schools across the country on the day after the Pearl Harbor attack, girls and boys talked about the war. "Fear of what was to come permeated the air," recalled a homefront child. Naturally the children had many questions: "Would the Japs come over here and bomb us next?" "Where would they bomb if they came?" "Would they come *even before Christmas?*" "What are we going to do?" asked a ten-year-old girl in rural Indiana. "I am afraid."[4]

Soon air raid drills started in the schools. Some children panicked. Hearing the sirens, one girl marched with her classmates to the basement and sat against the wall. "Right at that time, I heard an airplane and started to scream: I knew for sure we were going to be bombed." Even though these drills occurred during daylight and not darkness, some of the children were terrified because they were separated from their mothers and fathers. "I was always afraid," confessed a homefront girl, "that we would get bombed when I was at school and not at home with my family. That terror stuck with me for many, many years." And this was just the point. "Duck and cover," the civil defense drill introduced in the early 1950s, was nothing new to these homefront veterans of the war. Not only in civil defense but in other important ways, they had been programmed for the nuclear age.[5] In the 1940s there also was continuity between the war games children played during World War II and those they replayed during the Cold War. Some of these games had been around for generations but in the guise of cowboys and Indians or cops and robbers. "Red Fascism" did call for a change in the enemy's identity—if not in his essential characteristics—but the rules and practices of children's warfare remained the same. So too did the sex-typing that pervaded the battlefield: boys were the heroic soldiers and pilots, while girls were enemy soldiers, prisoners of war or, most often, nurses.[6] For years after the war, children continued not only to fight war games on school playgrounds and to dig foxholes in vacant lots, but to chant the same rhymes:

> Whistle while you work,
> Hitler is a jerk,
> Mussolini is a weeny,
> And Tojo is a jerk.[7]

Another point of continuity in the 1940s was children's popular culture, including radio, comic books, and particularly the movies, which changed little after the war. This was the last radio or at least pre-television generation. In 1946 almost 34 million households had radio sets, while only eight thousand owned televisions. By 1950 the figures stood at 41 million radio households and under four million television households. This jump in televisions was sizable, but in the next few years the increase was truly astronomical, to almost 44 million television households in 1959. For most American children, the 1950s was the first television decade.[8]

In the 1940s, however, most American children still listened to the radio. Their crime-stopping, spy-busting, battlefield heroes featured Superman, Captain Marvel, Jack Armstrong, and Don Winslow of the navy.[9] During the 1940s, particularly during the war years, homefront girls and boys remembered listening in silence to the news on the radio as their parents either talked among themselves or answered the children's questions. With television, however, the parents ceased to be the main interpreters of outside events. As the anthropologist Margaret Mead observed, "for the first time the young are seeing history made *before* it is censored by their elders."[10]

One wonders about the impact of television on those boys and girls who came of age after World War II. Was their childhood different from that of the homefront children? One answer has been ventured by a psychologist, Joshua Meyrowitz, who contends that the television years "have seen a remarkable change in the image and roles of children. Childhood as a protected and sheltered period of life has all but disappeared. . . . In the shared environment of television, children and adults know a great deal about one another's behavior and social knowledge—too much, in fact, for them to play out the traditional complementary roles of innocence versus omniscience."[11]

Meyrowitz, however, is a better psychologist than he is a historian, for even without television American childhood during the 1940s was anything but innocent. First, children listening to radio generated their own visual images; research done after the war concluded that radio stimulated the child's imagination "significantly more" than television did.[12] Second, the children did have access—direct and immediate access—to powerful external visual images, in which no adult stood between image and child. One example is photographs of the war, especially in the pages of *Life* magazine; another example is the newsreels; and a third is the war films made by Hollywood.

The impact of photographs could be shattering. *Life*'s photographs, for example, compelled numerous girls and boys not only to reorder their sense of what was right and wrong, but also to recalculate the outside limits of human horror. "*Life* did more than anything else," wrote a homefront boy in West Virginia, "to give me an adult's perspective of the war." Decades after the war, the images of *Life*'s photographs—"fiercely gruesome photos," one homefront girl called them—still evoke deep emotion. For one girl, it was a photograph "that showed a very young boy standing on his stumps . . . that brought home to me the horrors of war." For another, it was a picture "of a Japanese officer beheading a blindfolded American flyer. I see it in my mind very clearly," she wrote forty-seven years later.[13]

It was *Life*'s photographs of the Holocaust that most deeply moved the homefront children. "Most of my life," wrote a homefront girl, "I have been haunted by a photograph from *Life* magazine." It showed a young woman "standing in a Betty Grable–like pose, looking over her shoulder. Her skirt was raised, rather like a 'pin-up girl,' but, instead, she was displaying horribly disfigured legs, the result of medical experimentation." For another girl, the Holocaust photographs "exceeded anything we could have imagined—even in our wildest nightmares." Susan Sontag, the writer, was a twelve-year-old when she saw "photographs of Bergen-Belsen and Dachau which I came across by chance in a bookstore in Santa Monica in July 1945. Nothing I have seen—in photographs or in real life—ever cut me so deeply, instantaneously. Indeed, it seems plausible to me to divide my life into two parts, before I saw those photographs (I was twelve) and after. . . . When I looked at those photographs, something broke. Some limit had been reached, and not only that of horror; I felt irrevocably grieved, wounded, but a part of my feelings started to tighten; something went dead; something is still crying."[14]

A second major source of uncensored visual images was the newsreels, which one reporter called "a sort of *Life* magazine made animate." The newsreels were "our only view of the war," recalled a homefront girl. Not only did they bring "The Eyes and Ears of the World" to the audience, but they made "the war become very real." In black and white, the children saw scenes from air, land, and sea battles. During the war, about three-fourths of the newsreels showed military or naval hostilities or war-related activities.[15] What the children saw shocked them in ways that often persisted for years.

"The most terrifying thing I can remember," wrote a homefront girl, "was our weekly walk to the movie theater." She was afraid of what she would soon see, the newsreels, which gave her "terrible nightmares. It was apparent even to a child," she remembered, "that this was not part of a movie but reality." Moreover, the audience was not passive; this was an interactive medium. When a newsreel came on, she wrote, some people cried, others shouted at the images of Hitler and the Japanese, and "some were terrified—as I was."[16]

A final source of visuals was the feature war films made in Hollywood. Nearly one-third of the films made in 1943, at the height of the war, dealt with the war, and the messages of most of these were simplistic. According to film historians Albert Auster and Leonard Quart, Hollywood's war movies accorded "with the public's view that the war was a just and necessary act, fought against an evil enemy who threatened civilization." Hollywood "expressed little or no ambiguity about the reasons for fighting the war, the nature of the enemy, or life on the front itself." The films also "affirmed a faith in the American Way of Life—in our warmth, our honor, and our egalitarianism." In reply to questions about what it was that the United States was fighting for, the answer in several wartime movies was "pumpkin pie!"

As an aviator in *Thirty Seconds over Tokyo* (1944) put it, "When it's all over . . . just think . . . being able to settle down . . . and never be in doubt about anything." Just as the films indulged in both blatant and subtle sex-typing, so too they spun myths about American society. "One big family—that is America," stated one film character. "We all see eye to eye," asserted another.[17]

For America's children, these attitudes and values were perpetuated by the movies made after 1945 about the Second World War. Thomas Doherty, in his study of Hollywood and the cultural legacy of the war, has observed that any examination of "Hollywood and American culture since 1945 reveals an enduring, dynamic presence for the cinema of, and about, the Second World War."

"More than any other war," Doherty explains, "more than any other twentieth-century experience, it was motion picture friendly." Part of the war's "magnetic pull" was that it offered "the serenity of moral certainty. For Hollywood and American culture the Second World War would always be a safe berth."[18]

When the war ended, one might have expected a cessation of war films. But as the literary critic Alfred Kazin has written, psychologically World War II was a "war that has never ended." For the next twenty years, combat films of the

Second World War were a Hollywood staple.[19] And who would constitute the postwar audience for these films? According to a wartime prediction in the *New York Times,* it would be "the boys back from the battle [who] will love the war pictures and eat them up." But, as Doherty has explained, "most of the boys back from battle retired to lawnmowers and television. It was the boys of the boys back from battle who became the most enthusiastic and dedicated audience for motion pictures of World War II."[20]

It was not just the veterans' sons but also their daughters who were avid moviegoers. Without regard to gender, film images of the war filled the minds of American children in the 1940s. Yet the messages in these movies, as well as those conveyed in the radio shows, comic books, and war games, were gender-specific; the men fought the war, the women waited. Forty years after the war, a homefront girl, the political philosopher Jean Bethke Elshtain, wrote that her story was "not a soldier's story; it is, instead, a tale of how war, rumors of war, and images of violence, individual and collective, permeated the thoughts of a girl growing up . . . in the 1950s."

"Born in 1941," she wrote, "I knew the war only at second hand." Her encounter "as a child and citizen-to-be with the larger, adult world of war and collective violence" was, she observed, "filtered down to me through movies." Also through these films, she and countless others, boys as well as girls, learned lessons—and myths—and developed idealized, sex-typed notions, some of which in fact had been around for ages in epic tales of good versus evil. And in the absence of new story lines, Elshtain recalled, the film images of World War II told of "the Just Warrior and the Beautiful Soul."[21]

Not surprisingly, as one homefront boy pointed out, "a lot of the perspective on the war was taught to the children . . . after the fact through the movies made *after* the war." It was evident by the late 1940s that films about the Second World War were more entrenched than ever. "World War II Pix Kayo . . . Pacifism with $25,000,000 Gross in 1949," proclaimed a headline in *Variety;* "war proved no box office poison," read the front-page story. 1949 was a vintage year, including *Command Decision, Task Force, Battleground,* and the classic *Sands of Iwo Jima.*[22]

In addition to the Holocaust, the atomic bomb entered the consciousness of America's children at the end of World War II. In 1945 few American children expressed opposition to dropping the bomb. As far as they were concerned, it had to be done to win the war. "The bomb saved my brother's life," remembered a homefront girl."[23] But while America's homefront children accepted the necessity of using the atomic bomb, they nonetheless felt terror when they contemplated their own deaths in an all-out nuclear attack. "Packaged Devastation," a

November 1945 article in *Boys' Life* called it. A homefront boy recalled that "fear, bewilderment, and confusion swept the minds of grade schoolers when they heard the news." Once they saw newsreels of the bomb's explosion, a new horror entered their lives.

"Every newsreel," wrote a homefront girl, "showed the mushroom cloud over and over," and *Life*'s photos were also "imprinted in my mind's eye forever." For her the message was clear: "There is no place to hide."[24]

We have learned a great deal about the culture of the atomic age from historians, particularly Paul Boyer, Spencer R. Weart, Allan M. Winkler, and Margot A. Henriksen. "Fear psychosis" gripped many Americans, including boys and girls, in the months following Hiroshima and Nagasaki. People read that any nation could build an atomic bomb in five years, and they felt powerless. They wondered what would happen if their cities were bombed. Boyer had written that a "primal fear of extinction cut across all political and ideological lines"; Americans were not only frightened, but confused, not knowing how to respond to this "psychic event of almost unprecedented proportions."[25]

Gradually children's nuclear fears subsided, but such fears never totally disappeared. In August 1946, for example, the War Department released newsreels of the atomic bomb test at Bikini Atoll (Operation Crossroads), featuring "the grotesque beauty of the mushroom cloud," as well as footage of Hiroshima shot by Japanese cameramen right after the bombing. The victims' skin was seared; a man missing an ear appeared, then staggered on, as did a woman whose body was imprinted with the design of the dress she had been wearing. And in the fall of 1949, when the Russians exploded their first atomic bomb, the nuclear specter loomed again. In the following months, several events signaled the change in children's lives: the initiation of atomic air raid drills in various cities in 1950; the issuance of metal identification tags; the establishment of the Federal Civil Defense Administration in 1951; and in the same year, the release of the "Bert the Turtle says 'Duck and Cover'" animated film and the initiation of nationwide drills.

The decade of the 1950s saw not only Joseph McCarthy and the Korean War, but also concern over radioactive fallout (strontium 90 in milk) and, beginning in the later 1950s, a craze to build bomb shelters. And monsters—atomic mutants—began to inhabit children's comic books as well as the movies they saw. Gigantic ants appeared in *Them!* (1954) and a gigantic octopus in *It Came from Beneath the Sea* (1955). Watching *Them!* in the theaters, the children squirmed as they saw soldiers, with flamethrowers, fighting the giant ants in the sewers of Los Angeles—yet the ants kept coming. But sometimes the atomic contamination made people smaller, as in *The Incredible Shrinking Man* (1957);

in this movie, the hero wondered whether he had become "the man of the future" and whether he would be joined by other radiation victims in "this vast new land." Whether in films, books, or conferences, Paul Boyer has written, "nuclear fear was a shaping cultural force in these years." No one, it is safe to say, was more impressionable than the children.[26]

Because of an absence of research on children, we will never know how fearful America's children were of the nuclear specter in the 1940s and 1950s. Serious research on this topic did not begin until the early 1960s, when Milton Schwebel and Sibylle Escalona investigated children's feelings about the threat of nuclear war. "I was rather astounded by the intensity of their feelings," Schwebel reported. "It was something that wasn't known to us. Not only was it not discussed in the schools, but I surmised it wasn't discussed at home."[27]

There is anecdotal and even literary evidence of the intensity of children's fears. In 1947 a boy wrote a poem published in *Senior Scholastic:*

> I am a child of the Atom
> And I must make my roots in this age.
> But I am afraid
> And I would go back into my mother's
> womb where it is dark and quiet. . . .[28]

So far, this portrait of children's lives in postwar America has been most distressing. Clearly there also was a highly positive side, which included affluence, new homes and schools in the suburbs, and a proliferation of toys and fads. In the 1940s, too, Dr. Spock finally supplanted Dr. Watson, thus liberalizing not only child-rearing ideologies, but—if we are to believe contemporary psychological studies—child-rearing practices as well.

In a 1946 article entitled "Then and Now," Clara Savage Littledale, editor of *Parents' Magazine,* recounted twenty years of changes in the magazine's child-rearing advice to parents. Upon its founding in 1926, she wrote, the magazine "rode on the wave of a tremendous concern with [properly] bringing up children," and it expressed this concern by championing the rigidly behavioristic teachings of John B. Watson. Watson urged "habit training" in feeding, toilet training, and every other aspect of an infant's development. Such training, Littledale wrote, "should begin as soon as the child was born. The idea was that if you caught him early enough and trained and trained, allowing no deviations from the ideal but everlastingly hauling him up in the way he should go, why, you would never have any trouble at all. How could you, if you did the right thing every minute?" That was "then," twenty years ago, but in terms of current child-rearing philosophy, it seemed lights years ago. For "now" was 1946, the year when *The Common Sense Book of Baby and Child Care* by Dr. Benjamin Spock made its appearance.[29]

As cold as Watson's methods were, they commanded obedience from child-rearing experts as well as numerous American mothers.[30] Watson directed mothers to ignore baby's demands: "No feeding five minutes before schedule because a baby cried, chewed his fists and gave every indication of hunger. No feeding five minutes late—that was enough to prove any woman a Bad Mother. Right on the dot it must be." The same rigidity applied to toilet training; and as for affection, there should be little. "If the baby cried—let him cry! All babies cried. . . . But by all means don't pick him up."[31] Watson did recommend two forms of familiarity, one for bedtime, the other for first thing in the morning. "If you must," he wrote in 1928, "kiss them once on the forehead when they say goodnight. Shake hands with them in the morning."[32]

Watson's child-rearing advice had become the conventional wisdom of middle-class parents by the early 1930s, but his influence was soon under challenge.[33] Many factors—some professional, some theoretical, some ideological—conspired to undermine Watson's primacy. Academically, research in child development flourished in the 1930s at various universities, including Iowa, Johns Hopkins, Minnesota, Yale, Teachers College, and California.[34] At these schools such experts as Dr. Adolph Meyer (director of the psychiatric clinic at Johns Hopkins University), Dr. C. Andrews Aldrich (a pediatrician at the University of Minnesota), and coauthor Mary M. Aldrich offered alternatives to Watson. Meyer urged parents to develop a better understanding of how their children grew emotionally as well as physically; in 1938 the Aldriches' book *Babies Are Human Beings* was published, admonishing parents to respond to their babies' needs. The Aldriches warned that most of our "spoiled children are those who, as babies, have been denied essential gratifications in a mistaken attempt to fit them into a rigid regime."[35]

In addition, in the United States Freudian theory had begun to exert a weighty influence on theories of child development and parent-child relations. Psychoanalysts focused therapeutic attention on breast feeding and toilet training as being bound up with psychosexual development and with underlying anxieties and guilt, and they argued that forced training according to the Watsonian paradigm could produce trauma and lifelong damage to the individual. Drawing upon Freudian theory, American child-rearing experts took cognizance of infant sexuality, the unconscious, and the oedipal conflict. Instead of recommending "tight scheduling," they urged mothers to trust their own judgment.[36] Another significant influence was Dr. Arnold Gesell, a pediatrician at Yale, who countered Watson's extreme environmentalism with his emphasis on physical maturation as the primary determinant of child development. In

1940 Gesell advised mothers and fathers that "the child is in league with nature and he must do his own growing." Finally, during the war years, Benjamin Spock was beginning to publish portions of *The Common Sense Book of Baby and Child Care,* such as his March 1945 article in *Parents' Magazine,* "A Babies' Doctor Advises on Toilet Training."[37]

World War II was a crucial reason for the transition in American child rearing from parent-centered scheduling to child-centered permissiveness. Politicians and parents alike worried about children who had been buffeted first by the Great Depression and then by the Second World War. Believing that children needed to be held in troubled times, more and more parents sought to redeem "the promise of the 'secure' child in a 'world of change.'" Regimentation, on the other hand, smacked of fascism.[38]

At the same time, child-rearing experts, politicians, and parents all expressed alarm that the wartime disruption of family life had fostered "conditions" which, according to the writer Vance Packard, "could hardly be worse for baby raising." Fathers serving at distant duty stations in the armed forces, mothers working long hours in factories and shipyards, "latchkey children" at home alone—"all these handicaps," Packard wrote, "imposed by war and modern conditions that I've mentioned add up to a lot of perplexity for a growing child." Packard's advice: "Give the war babies a break."[39]

So pronounced was the shift in American child-rearing advice during the war that one sociologist has declared that "the decade from 1935 to 1945 may well be called 'baby's decade' when mother becomes secondary to . . . the baby's demands." It is clear that the trend toward "self-demand babies" accelerated during World War II. One can compare the 1938 and 1942 editions of *Infant Care* to appreciate the change. In contrast to the 1938 edition's worried view of babies' autoerotic "impulses," the 1942 manual considered these drives to be weak and almost inconsequential. *Infant Care* viewed babies in 1942 as explorers who "want to handle and investigate everything that they can see and reach. When a baby discovers his genital organs he will play with them." But this was no cause for mother to tie baby's feet apart or glove his tiny hands. "See that he has a toy to play with and he will not need to use his body as a plaything." Most important, parents should attend to their babies' "wants" as well as their "needs." Babies cried for attention, but that was all right. "Babies want attention; they probably need plenty of it." Similarly, parents should play with their babies; amusement and fun, which earlier editions of *Infant Care* had disdained as titillating and excessively exciting, should characterize parent-child relations. "Play and singing make both mother and baby enjoy the routine of life," stated the 1942 edition.[40]

America's homefront boys and girls thus were at the center of an important transition. Even by 1940, child-rearing advice had changed considerably. Now only one-third of the articles recommended the behavioristic regimen, while two-thirds urged self-regulatory, "permissive" approaches. In contrast to the behaviorists' strictly enforced feeding schedules, Josephine Kenyon, writing in *Good Housekeeping,* said simply: "It is reasonable to feed a baby when he's hungry. . . . It is unreasonable to make him wait." The balance had shifted from "tight scheduling" to "self-regulation," and by 1948 the shift was complete; none of the articles published that year advocated Watsonian methods.[41]

Also during wartime, the child-rearing experts urged mothers to recognize that their children were resilient. In mid-1942 P. L. Travers, the author of the Mary Poppins books, wrote in the *Woman's Home Companion* that "children are tough individuals, much tougher than grown-ups when it comes to facing facts."[42] All the experts agreed that parents, particularly mothers, should be honest with their daughters and sons, letting them "know what to expect," but still seeing to it "that childhood has its full measure of happiness" and responding to emergencies "with cheerful competence." It was imperative that children, "too, in the midst of danger, shall have courage." At the same time, parents should protect their sons and daughters from hatred. "Perhaps out of this war," stated *Parents' Magazine,* "will come a better world for our children. Surely it will come quicker if we can save them from hatred and fear and bitterness."[43]

This deep and widely shared concern about children's welfare was good news for America's children not only during the war but throughout the decade of the 1940s. Almost without exception, the wartime advice began by reminding parents that "babies need mothering, the good old-fashioned kind that goes with rocking and singing lullabies." Children needed love and affection. Victory in warfare would be hollow indeed if the homefront children should be emotionally stunted.

The psychologist Frances Degen Horowitz has hypothesized that "one might look upon the Spock era . . . as a kind of reaffirmation of democratic principles taken down to child rights in a reaction against fascism and authoritarianism as they were thought to have produced obedient but morally manipulable children."[44]

Child-centered approaches began to dominate the advice columns. One of the strongest statements was *The Rights of Infants* (1943) by Dr. Margaret A. Ribble, who recommended feeding and bathing as times for talk and loving play. Her prescription for the baby's emotional feeding was "a little at a time, and frequently." Spurred by the war's liberal ideology, an important change was taking place. Just as democracy had prevailed over authoritarianism abroad, so too democratic notions had begun to transform child rearing in America. "Belief in our children is the essence of democracy, of what we call our way of life," wrote the vice president of the Child Study Association of America in 1944. "It is to preserve our way of life that we are fighting this war."[45]

Published in 1946 was Dr. Spock's *Baby and Child Care,* which was the culmination of a ten-year process of revisionism—and went on to sell 22,000,000 copies over the next twenty-five years. Spock's guide appeared as the shift toward liberalized child rearing was gaining wide acceptance, and while this book accentuated the movement, it certainly did not create it.[46] In fact, various investigators have reported that American mothers—regardless of social class—became markedly less coercive throughout the 1940s. Based on "observed maternal behavior," two such researchers, Elinor Waters and Vaughn J. Crandall, affirmed that babying, protectiveness, and affection peaked in 1950, the height of the "permissive era."[47]

While by 1950 the pendulum in published child-rearing advice had swung to permissiveness, the question remains: did parents actually heed the new advice? Urie Bronfenbrenner, the psychologist, has contended that middle-class parents in the 1940s and 1950s did. Child-rearing "practices are likely to change most quickly," Bronfenbrenner has written, "in those segments of society which have closest access [to] and are most receptive to the agencies or agents of change (e.g., public media, clinics, physicians, and counselors)."[48] Clearly, however, the extent to which child-rearing advice actually penetrated the consciousness of white-collar, middle-class America is still at issue. So too is the extent to which the working class adopted permissive approaches to child rearing.[49] But considering the proliferation of books and articles advocating permissiveness in child rearing—not to mention the wisdom contained in many of these pieces—one can, I think, reasonably assume that the advice reached many eager readers. And much social-science evidence exists to show that both the middle and working classes became more permissive in the 1940s and 1950s.[50]

Historians agree that a distinctive era in the history of the United States—indeed, of the world—began in 1945, with the end of World War II and the advent of the atomic age and the Cold War. In America in 1945, historians explain, after a decade of economic depression and almost four years of war, there was a new era of abundance and family togetherness. There was a new enemy as well as a new all-powerful weapon. The historians are correct in identifying 1945 as a pivotal year in history. But if in contextualizing children's lives in postwar America we looked only at postwar events, we would fail to see the continuities from the war years, particularly in children's popular culture and in their secret fears of death and destruction. After all, these were the girls and boys who, by the millions and in numerous ways, had fought the Second World War on the homefront. They were already veterans by the time they had to confront the postwar events of the Cold War and the atomic age. And if we focused only on permissive child rearing in the aftermath of Dr. Spock's best-seller published in 1946, we would miss the crucial years during the war for the development of liberalized child rearing in the United States.

During the Second World War, America's children were patriotic to the core; in the years after 1945, they were ready—emotionally, cognitively, ideologically—to continue doing their duty during the Cold War. Yet in many ways the Cold War was a strange time to be a child. Children in postwar America were perplexed, and their lives were filled with a basic contradiction. In an era of permissive child rearing, their lives seemed especially precious to their parents and to society generally, and yet, as the children themselves knew, they could be incinerated in a split second. This was a prescription if not for schizophrenia, then at least for deep ambivalence about life and the future.

Acknowledgments

I would like to thank David M. Katzman, Barry Shank, and Allen M. Winkler for their helpful comments on my chapter.

Notes

1. John Higham, *Strangers in the Land: Patterns of American Nativism, 1860–1925* (New York: Atheneum, 1963), 223; Robert K. Murray, *Red Scare: A Study in National Hysteria, 1919–1920* (Minneapolis: University of Minnesota Press, 1955); William Preston Jr., *Aliens and Dissenters: Federal Suppression of Radicals, 1903–1933* (Cambridge, Mass.: Harvard University Press, 1963); *The Fleet Review* 10 (July 1919), 17.
2. Les K. Adler and Thomas G. Paterson, "Red Fascism: The Merger of Nazi Germany and Soviet Russia in the American Image of Totalitarianism, 1930s–1950s," *American Historical Review* 75 (April 1970), 1046–64. Whether called "containment culture," "victory culture," or "Dr. Strangelove's America," the culture of the Cold War has been studied from various perspectives, among them liberalism, radicalism, and postmodernism. Among the best of these studies are Stephen J. Whitfield, *The Culture of the Cold War* (rev. ed., Baltimore: Johns Hopkins University Press, 1996); Lary May, ed., *Recasting America: Culture and Politics in the Age of the Cold War* (Chicago: University of Chicago Press, 1989); Alan Nadel, *Containment Culture: American Narratives, Postmodernism, and the Atomic Age* (Durham, N.C.: Duke University Press, 1995); Margot A. Henriksen, *Dr. Strangelove's America: Society and Culture in the Atomic Age* (Berkeley: University of California Press, 1997); "Special

Section: The Politics of Culture in the Cold War," *Prospects* 20 (1995), 449–541. Most studies pay inadequate attention to cultural continuities in the 1940s; an exception, which delves into the long shadow cast by the Second World War over the culture of the Cold War, is Tom Engelhardt, *The End of Victory Culture: Cold War America and the Disillusioning of a Generation* (New York: Basic Books, 1995).

3. Daniel Beekman, *The Mechanical Baby: A Popular History of the Theory and Practice of Child Rearing* (Westport, Conn.: Lawrence Hill, 1977), 163–65.
4. Letters 136, 191, 270, 342; Lauretta Bender and John Frosch, "Children's Reactions to the War," *American Journal of Orthopsychiatry* 12 (October 1942), 573, 579–80./ These letter numbers refer to the letters which were written to me in 1990 and 1991 by American homefront children, that is, women and men born between 1933 and 1945. I had written the hundred largest newspapers in the country and about forty magazines asking the editors to let their readers know of my project and asking homefront children to write me with their memories of the war. Many did; in all, I received 2500 letters dealing with a wide range of topics. These letters form an important component of my book *"Daddy's Gone to War": The Second World War in the Lives of America's Children* (New York: Oxford University Press, 1993).
5. Letters 21, 28, 25, 63, 186, 240C, 244, 355; Elisabeth R. Geleerd, "Psychiatric Care of Children in Wartime," *American Journal of Orthopsychiatry* 12 (October 1942), 589–91; Mildred Burgum, "The Fear of Explosion," *American Journal of Orthopsychiatry* 14 (April 1944), 349–57; JoAnne Brown, "'*A* Is for Atom, *B* Is for Bomb': Civil Defense in American Public Education, 1948–1963," *Journal of American History* 75 (June 1988), 84.
6. Tuttle, *"Daddy's Gone to War,"* 134–47.
7. Letters 32, 137, 174, 413, 442D.
8. U.S. Bureau of the Census, *Historical Statistics of the United States, Colonial Times to 1970, Bicentennial Edition* (Washington, D.C.: Government Printing Office, 1975), Part 2, 796.
9. Tuttle, *"Daddy's Gone to War,"* 148–51.
10. Mead is quoted in *Newsweek* 92 (November 27, 1978), 75–76. Books that deal with the harmful effects of television on children's development include Kenneth Keniston and the Carnegie Council on Children, *All Our Children: The American Family under Pressure* (New York: Harcourt Brace Jovanovich, 1977); David Elkind, *The Hurried Child: Growing Up Too Fast Too Soon* (Reading, Mass.: Addison-Wesley, 1981); Valerie Polakow Suransky, *The Erosion of Childhood* (Chicago: University of Chicago Press, 1982); Neil Postman, *The Disappearance of Childhood* (New York: Delacorte, 1982); Marie Winn, *Children without Childhood* (New York: Pantheon, 1983); Vance Packard, *Our Endangered Children: Growing Up in a Changing World* (Boston: Little, Brown, 1983); Richard Louv, *Childhood's Future: New Hope for the American Family* (Boston: Houghton Mifflin, 1990). A valuable report sponsored by the American Psychological Association is Aletha C. Huston et al., *Big World, Small Screen: The Role of Television in American Society* (Lincoln: University of Nebraska

Press, 1992), which summarizes numerous psychological studies and draws signficant conclusions about television's role in American life.

11. Joshua Meyrowitz, "The Adultlike Child and the Childlike Adult: Socialization in an Electronic Age," *Daedalus* 113 (Summer 1984), 19–48; Joshua Meyrowitz, *No Sense of Place: The Impact of Electronic Media on Social Behavior* (New York: Oxford University Press, 1985), 226–67.
12. Marie Winn, "Why Has Radio Tuned Out Children?" *New York Times*, September 25, 1983.
13. Letters 12B, 187, 276, 316, 405.
14. Letters 97E, 442B.
15. Letters 150, 312, 318, 350, 435; Raymond Fielding, *The American Newsreel, 1911–1967* (Norman: University of Oklahoma Press, 1972), 288–95; *Kansas City Star*, April 21, 1986; Roger Manvell, *Films and the Second World War* (South Brunswick, N.J.: A. S. Barnes, 1974), 122–23, 183–85; George H. Roeder Jr., *The Censored War: American Visual Experience during World War II* (New Haven: Yale University Press, 1993).
16. Letters 42, 150, 222; Fielding, *The American Newsreel*, 295.
17. Auster and Quart, *How the War Was Remembered: Hollywood and Vietnam* (New York: Praeger, 1988), 6; Dorothy B. Jones, "The Hollywood War Film: 1942–1944," *Hollywood Quarterly* 1 (1945), 3; Jackson Lears, "A Matter of Taste: Cultural Hegemony in a Mass-Consumption Society," in Lary May, ed., *Recasting America: Culture and Politics in the Age of the Cold War* (Chicago: University of Chicago Press, 1989), 41; Bernard F. Dick, *The Star-Spangled Screen: The American World War II Film* (Lexington: University of Kentucky Press, 1985), 174–87, 324–27; Leo A. Handel, *Hollywood Looks at Its Audience: A Report of Film Audience Research* (Urbana: University of Illinois Press, 1950), 120–24; Lawrence W. Levine *Highbrow/Lowbrow: The Emergence of Cultural Hierarchy in America* (Cambridge, Mass.: Harvard University Press, 1988), 243–56. For sex-typing in wartime films, see Kathryn Weibel, *Mirror Mirror: Images of Women Reflected in Popular Culture* (Garden City, N.Y.: Anchor, 1977), 116–19; M. Joyce Baker, *Images of Women in Film: The War Years, 1941–1945* (Ann Arbor, Mich.: UMI Research Press, 1981); Michael Renov, *Hollywood's Wartime Woman: Representations and Ideology* (Ann Arbor: UMI Research Press, 1988). Important studies of films of the 1940s are Carol Trayner Williams, *The Dream Beside Me: The Movies and the Children of the Forties* (Rutherford, N.J.: Fairleigh Dickinson University Press, 1980), 41–80; Barbara Deming, *Running Away from Myself: A Dream Portrait of America Drawn from the Films of the Forties* (New York: Grossman, 1969), 8–10; Dana Polan, *Power and Paranoia: History, Narrative, and the Cinema, 1940–1950* (New York: Columbia University Press, 1986); Martha Wolfenstein and Nathan Leites, *Movies: A Psychological Study* (Glencoe, Ill.: Free Press, 1950). For the war films, see Colin Shindler, *Hollywood Goes to War: Films and American Society, 1939–1952* (London: Routledge & Kegan Paul, 1979); Roger Manvell, *Films and the Second World War* (South Brunswick, N.J.: A. S. Barnes, 1974); Clayton R. Koppes and Gregory D. Black, *Hollywood Goes to War: How Politics,*

Profits, and Propaganda Shaped World War II Movies (New York: Free Press, 1987); Jeanine Basinger, *The World War II Combat Film* (New York: Columbia University Press, 1986); Kathryn Kane, *Visions of War: Hollywood Combat Films of World War II* (Ann Arbor, Mich.: UMI Research Press, 1981); Allen L. Woll, *The Hollywood Musical Goes to War* (Chicago: Nelson-Hall, 1983); K. R. M. Short, ed., *Film and Radio Propaganda in World War II* (London: Croon Helm, 1983); Peter A. Soderbergh, "The Grand Illusion: Hollywood and World War II, 1930–1945," *University of Dayton Review* 5 (Winter 1968–1969), 13–21; Peter A. Soderbergh, "The War Films," *Discourse* 11 (Winter 1968), 87–91; Peter A. Soderbergh, "On War and the Movies: A Reappraisal," *Centennial Review* 11 (Summer 1967), 405–18; Lewis Jacobs, "World War II and the American Film," *Cinema Journal* 7 (Winter 1967–1968), 1–21.

18. Doherty, *Projections of War: Hollywood, American Culture, and World War II* (New York: Columbia University Press, 1993), 265–72.
19. Alfred Kazin, *Bright Book of Life* (Boston: Little, Brown, 1973), 81; Auster and Quart, *How the War Was Remembered,* 8; Doherty, *Projections of War,* 272; Ball, "The Politics of Social Science in Postwar America," in May, *Recasting America,* 82.
20. Doherty, *Projections of War,* 271.
21. J. B. Elshtain, *Women and War* (New York: Basic Books, 1987), xii, 14–17; Walter Goodman, "Romance Narrows the Generation Gap," *New York Times,* August 10, 1986.
22. Letters 97E, 119, 120; Linda Dittmar and Gene Michaud, "America's Vietnam War Films: Marching Toward Denial," in Dittmar and Michaud, eds., *From Hanoi to Hollywood: The Vietnam War in American Film* (New Brunswick, N.J.: Rutgers University Press, 1990), 4–6; "Refugio," *U.S. News & World Report* 107 (December 18, 1989), 51; Basinger, *World War II Combat Film,* 296.
23. Letters 16, 102, 408.
24. Burr Leyson, "Packaged Devastation," *Boys' Life* (November 1945), 3, 40–42; Letters 28, 97E 184, 206P, 264, 446.
25. Boyer, *By the Bomb's Early Light;: American Thought and Culture at the Dawn of the Atomic Age* (New York: Pantheon Books, 1985), xvii, 15, 21–26, 32; Weart, *Nuclear Fear: A History of Images* (Cambridge, Mass.: Harvard University Press, 1988), 265–66; Winkler, *Life under a Cloud: American Anxiety about the Atom* (New York: Oxford University Press, 1993), 137; Leonard S. Cottrell Jr. and Sylvia Eberhart, *American Opinion on World Affairs in the Atomic Age* (Princeton: Princeton University Press, 1948), 15–29.
26. Boyer, *By the Bomb's Early Light,* 318, 328, 334–51, 354–55; Weart, *Nuclear Fear,* 213–14, 312–15, 368–69; Winkler, *Life under a Cloud,* 114–16; Winkler, "A 40-Year History of Civil Defense," *Bulletin of the Atomic Scientists* 40 (June–July 1984), 16–22; Brown, "'*A* Is for Atom, *B* Is for Bomb,'" 68–90; Michael J. Carey, "The Schools and Civil Defense: The Fifties Revisited," *Teachers College Record* 84 (Fall 1982), 115–27; Stephen Kneeshaw, "Hollywood and 'The Bomb,'" Organization of American Historians *Newsletter* 14 (May 1986), 9–10; Doherty, *Projections of War,* 280–81; Boyer, "From Activism to Apathy: The American People and Nuclear Weapons," *Journal of American History* 70 (March 1984), 823.

27. Phyllis La Farge, *The Strangelove Legacy: Children, Parents, and Teachers in the Nuclear Age* (New York: Harper & Row, 1987), 23–25; J. H. Elder, "A Summary of Research on Reactions of Children to Nuclear War," *American Journal of Orthopsychiatry* 35 (January 1965), 120–23.
28. Robert Thom, "Self-Analysis," *Senior Scholastic,* October 6, 1947; quoted in Michael Scheibach, "The Atomic Generation: Coming of Age in the Postwar Era, 1945–1955" (unpublished paper); Michael J. Carey, "Psychological Fallout: Interviews with Members of the Generation of the 1940s on the Effect of Nuclear Weapons on Their Lives," *Bulletin of Atomic Scientists* 38 (January 1982), 20–24.
29. Littledale, "Then and Now," *Parents' Magazine* 21 (October 1946), 127; Catherine Mackenzie, "Bringing Up Baby—Then and Now," *New York Times Magazine* (February 8, 1943), 14, 23; Terry Strathman, "From the Quotidian to the Utopian: Child- Rearing Literature in America, 1926–1946," *Berkeley Journal of Sociology* 29 (1984), 1–34.
30. There are numerous studies of American child-rearing during the 1940s; see Urie Bronfenbrenner, "Socialization and Social Class through Time and Space," in Eleanor E. Maccoby et al., eds., *Readings in Social Psychology* (3d ed., New York: Holt, 1958), 400–25; Elise P. Fishbein, "New Ideas in Child Development," *Hygeia* 25 (March 1947), 201–2; Martha Wolfenstein, "Trends in Infant Care," *American Journal of Orthopsychiatry* 23 (January 1953), 120–30; Wolfenstein, "Fun Morality: An Analysis of Recent American Child-training Literature," in Margaret Mead and Martha Wolfenstein, eds., *Childhood in Contemporary Cultures* (Chicago: University of Chicago Press, 1955), 168–78; Bronfenbrenner, "The Changing American Child—A Speculative Essay," *Journal of Social Issues* 17 (1961), 6–18; L. J. Borstelmann, "Changing Concepts of Child Rearing, 1920s–1950s: "Parental Research, Parental Guidance, Social Issues" (paper delivered at Southeastern Conference on Human Development, 1976); Nancy Pottishman Weiss, "Mother, the Invention of Necessity: Dr. Benjamin Spock's *Baby and Child Care,*" *American Quarterly* 29 (Winter 1977), 519–29; Weiss, "The Mother-Child Dyad Revisited: Perceptions of Mothers and Children in Twentieth-Century Child-Rearing Manuals," *Journal of Social Issues* 34 (Spring 1978), 31–36.; Duane F. Alwin, "From Obedience to Autonomy: Changes in Traits Desired in Children," *Public Opinion Quarterly* 52 (Spring 1988), 33–52.
31. Littledale, "Then and Now," 127, 128.
32. Watson, *Psychological Care of Infant and Child* (London: Allen & Unwin, 1928), 73.
33. Lois Barclay Murphy, "Cultural Factors in the Development of Children," *Childhood Education* 23 (October 1946), 53; Celia B. Stendler, "Sixty Years of Child Training Practices: Revolution in the Nursery," *Journal of Pediatrics* 36 (January 1950), 129–31; Clark E. Vincent, "Trends in Infant Care Ideas," *Child Development* 22 (September 1951), 199–207.
34. Hamilton Cravens, "Child Saving in the Age of Professionalism, 1915–1930"; and Leroy Ashby, "Partial Promises and Semi-Visible Youths: The Depression and World War II," both in Joseph M. Hawes and N. Ray Hiner, eds., *American Childhood: A Research Guide and Historical Handbook* (Westport, Conn.: Greenwood, 1985), 448–

49, 502; Robert R. Sears, *Your Ancients Revisited: A History of Child Development* (Chicago: University of Chicago Press, 1975), 19–21; Margo Horn, *Before It's Too Late: The Child Guidance Movement in the United States, 1922–1945* (Philadelphia: Temple University Press, 1989), 135–53.

35. Sears, *Your Ancients Revisited,* 43–45; Borstelmann, "Changing Concepts of Childrearing," 10–12; Littledale, "Then and Now," 130, 132; Ashby, "Partial Promises and Semi-Visible Youths," 502–3; Michael Gordon, *Infant Care* Revisited," *Journal of Marriage and the Family* 30 (November 1968), 578–79; Mary Cable, *The Little Darlings: A History of Child Rearing in America* (New York: Scribner's, 1975), 182–85; Theresa R. Richardson, *The Century of the Child: The Mental Hygiene Movement and Social Policy in the United States and Canada* (Albany: State University of New York Press, 1989), 129–70; Neil M. Cowan and Ruth Schwartz Cowan, *Our Parents' Lives: The Americanization of Eastern European Jews* (New York: Basic Books, 1989), 177–209.

36. John R. Seely, "The Americanization of the Unconscious," *Atlantic Monthly* 208 (July 1961), 68–72. For Freudian theory's influence on Spock, see A. Michael Sulman, "The Humanization of the American Child: Benjamin Spock as a Popularizer of Psychoanalytic Thought," *Journal of the History of the Behavioral Sciences* 9 (July 1973), 258–65; William G. Bach, "The Influence of Psychoanalytic Thought on Benjamin Spock's *Baby and Child Care,*" *Journal of the History of the Behavioral Sciences* 10 (January 1974), 91–94.

37. Fishbein, "New Ideas in Child Development," 224; Terry Strathman, "From the Quotidian to the Utopian," 6–8; Littledale, "Then and Now," 130, 132, 136; Borstelmann, "Changing Concepts of Childrearing," 8–10; Gesell et al. *The First Five Years of Life: A Guide to the Study of the Preschool Child* (New York: Harper, 1940), 6–9, 319–43; Len Chaloner, "Breast Feeding Means Happiness for Two," *Parents' Magazine* 19 (February 1944), 18–19; Herman N. Bundesen, "The Best-Fed Are Breast-Fed," *Ladies' Home Journal* 61 (July 1944), 128; Spock, "A Babies' Doctor Advises on Toilet Training," *Parents' Magazine* 20 (March 1945), 20.

38. Ashby, "Partial Promises and Semi-Visible Youth," 503; Borstelmann, "Changing Concepts of Childrearing," 12–14; Murphy, "Cultural Factors in the Development of Children," 53; Steven Mintz and Susan Kellogg, *Domestic Revolutions: A Social History of American Family Life* (New York: Free Press, 1988), 164–65.

39. Packard, "Give the War Babies a Break," *American Magazine* 139 (May 1945), 24–25ff. See also two articles by Gladys Denny Shultz: "If Daddy's Gone to War," *Better Homes & Gardens* 22 (October 1943), 14ff.; "Life without Father," *Better Homes & Gardens* 22 (June 1944), 22ff.

40. *Infant Care* (Washington, D.C.: Government Printing Office, 1942), 41, 59–60; *Infant Care* (Washington, D.C.: Government Printing Office, 1945), 52; Wolfenstein, "Fun Morality," 169–74.

41. Clark E. Vincent, "Trends in Infant Care Ideas," 205; Littledale, "Then and Now," 136; Stendler, "Sixty Years of Child Training Practices," 126, 131; Kenyon, "Less Rigid Schedules for Baby," *Good Housekeeping* 110 (1940), 92; Herman N. Bundesen,

"Bad Habits in Children," *Ladies' Home Journal* 58 (August 1941), 83–84; A. H. Maslow and I. Szilagyi-Kessler, "Security and Breast-feeding," *Journal of Abnormal and Social Psychology* 41 (January 1946), 83–85; Ralph H. Ojetmann et al., "A Functional Analysis of Child Development Materials in Current Newspapers and Magazines," *Child Development* 19 (March–June 1946), 76–92.

42. Travers, "Children Are Tough," *Woman's Home Companion* 69 (June 1942), 48.
43. Clara Savage Littledale, "Explaining the War to Our Children," *Parents' Magazine* 17 (February 1942), 17; Mary Shattuck Fisher, "What Shall We Tell Children about War?" *Journal of Home Economics* 34 (May 1942), 277–79; Evelyn Ardis Whitman, "Keeping Young Chins Up," *Parents' Magazine* 17 (September 1942), 29, 94–95; Anna W. M. Wolf, "War and Discipline," *Parents' Magazine* 17 (October 1942), 20–21ff.; Frederick H. Allen, "Can the Youngest Take It?" *Parents' Magazine* 17 (November 1942), 26–27ff.; Barbara Biber, "Childhood as Usual?" *Parents' Magazine* 18 (February 1943), 26–27ff.; James Benjamin, "Let Your Child Know What to Expect," *Parents' Magazine* 19 (November 1944), 20–21ff.
44. Fishbein, "New Ideas in Child Development," 224–25; Horowitz to the author, August 5, 1983.
45. "Lullabies and Play Needed by Infants," *Science News Letter* 45 (April 1, 1944), 216; Child Study Association of America, *The Child, the Family, the Community: A Classified Booklist* (New York: Child Study Association of America, undated), 14–20, 51–52.
46. Fishbein, "New Ideas in Child Development," 224–25; Mary Jo Bane, "A Review of Child Care Books," *Harvard Educational Review* 43 (November 1973), 669; Ashby, "Partial Promises and Semi-Visible Youths," 503; book reviews in *American Journal of Nursing* 42 (October 1942), 1222; 44 (February, June, October 1944), 195, 524, 1015; Michael Zuckerman, "Dr. Spock: The Confidence Man," in Charles E. Rosenberg, ed., *The Family in History* (Philadelphia: University of Pennsylvania Press, 1975), 182–84; William Graebner, "The Unstable World of Benjamin Spock: Social Engineering in a Democratic Culture, 1917–1950," *Journal of American History* 67 (December 1980), 612–29.
47. Waters and Crandall, "Social Class and Observed Maternal Behavior from 1940 to 1960," *Child Development* 35 (December 1964), 1021–32; E. H. Klatskin et al., "The Influence of Degree of Flexibility in Maternal Child Care Practices on Early Child Behavior," *American Journal of Orthopsychiatry* 26 (January 1956), 79–93.
48. Bronfenbrenner, "Socialization and Social Class through Time and Space," 411. In agreement is Martha Sturm White, "Social Class, Child-Rearing Practices, and Child Behavior," *American Sociological Review* 22 (December 1957), 704–12. For an excellent discussion of the psychological literature on social class and "the trend toward permissiveness" from 1940 to 1960, see Edward Zigler and Irvin L. Child, "Socialization," in Gardner Lindzey and Elliot Aronson, eds., *The Handbook of Social Psychology*, vol. 3: *The Individual in a Social Context* (2d ed., Reading, Mass.: Addison-Wesley, 1968–1969), 483–501.

49. What is the evidence that working-class families became more permissive after the war? According to Bronfenbrenner, during the twenty-five-year period between 1933 and 1958, "parent-child relationships in the middle class are consistently reported as more acceptant and equalitarian, while those in the working class are oriented toward maintaining order and obedience." Yet the situation was in flux, as Bronfenbrenner conceded. Research done in the late 1950s and early 1960s held that working-class child rearing was becoming more liberalized while, in the middle classes, the trend to permissiveness was slowing: Bronfenbrenner, "Socialization and Social Class through Time and Space," 425; Nathan Maccoby, "The Communication of Child-Rearing Advice to Parents," *Merrill-Palmer Quarterly* 7 (July 1961), 200. On social class and parenting, see the following by Melvin L. Kohn, "Social Class and Parental Values," *American Journal of Sociology* 64 (January 1959), 337–51; "Social Class and the Exercise of Parental Authority," *American Sociological Review* 24 (June 1959), 352–66; Bronfenbrenner with Eleanor E. Carroll, "Social Class and the Allocation of Parental Responsibilities," *Sociometry* 23 (December 1960), 372–92; "Social Class and Parent-Child Relationships," *American Journal of Sociology* 68 (January 1963), 471–80; Bronfenbrenner with Leonard Pearlin, "Social Class, Occupation, and Parental Values: A Cross-National Study," *American Sociological Review* 31 (August 1966), 466–79; *Class and Conformity: A Study in Values* (2d ed., Chicago: University of Chicago Press, 1977).
50. Orville G. Brim Jr., *Education for Child Rearing* (New York: Russell Sage Foundation, 1959), 55; Lillian Cukier Robbins, "The Accuracy of Parental Recall of Aspects of Child Development and Child-Rearing Practices," *Journal of Abnormal and Social Psychology* 66 (March 1963), 261–70; Marian Yarrow et al., "Reliability of Maternal Retrospection: A Preliminary Report," *Family Process* 3 (1964), 207–18; Jay Mechling, "Advice to Historians on Advice to Mothers," *Journal of Social History* 9 (Fall 1975), 44–63; Zuckerman, "Dr. Spock: The Confidence Man," 179–82; Eleanor E. Maccoby and John A. Martin, "Socialization in the Context of the Family: Parent-Child Interaction," in Paul H. Mussen, ed., *Handbook of Child Psychology,* vol. 4: *Socialization, Personality, and Social Development,* E. Mavis Hetherington, vol. ed. (New York: Wiley, 1983), 16–18.

CHAPTER 2

Cold War Workers, Cold War Communities

Ann Markusen

merica was a mature economy, and a booming one, in the 1950s. Large factories were flush with orders, employment expanded, and unions enjoyed popularity and success. In cities like Detroit and Chicago, all across the industrial heartland, memories of the difficulties of the Depression were dimmed by new-found prosperity. The fact that it was a peaceful rather than a wartime decade dismissed fears that only a command economy serving a warring state could ensure sufficient employment for Americans.

The economic progress of the decade turned out to be transitory. The quantitative boom based on pent-up demand from forced wartime savings and orders for machinery and equipment from war-depleted Japan, Korea, and Europe petered out by the end of the 1950s. Meanwhile a quiet but dramatic change had taken place in the qualitative nature of the American economy, especially in the institutions responsible for technical innovation and the character of new, leading-edge industries. The commitment to a Cold War, based on threat (not action) and on continually improved weapons of extraordinary power and mobility, commandeered the lion's share of the nation's research and development capability and fueled a new set of industries—aircraft, communications, electronics, and computing—that were to irrevocably alter the American economic, occupational, and regional landscape.

In this chapter, I argue that important increments to American occupational and residential structure were the products of the Cold War, set in place in the critical decade of the 1950s. Because of the unique character of Cold War, certain occupations within science and engineering grew rapidly and became prestigious—nuclear physics, aeronautical engineering, oceanography. Simultaneously, technologies developed for Cold War purposes hastened the displacement of blue-collar workers in traditional manufacturing. Hosting new Cold War offices, plants, and facilities, new relatively segregated and anticosmopolitan regions and communities grew up in tandem: Orange Country and Colorado Springs are prototypical. While these occupational strata and the suburban communities that housed them were modestly sized in terms of aggregate employment and population, they loomed large in the incremental growth of the nation, and here I can only speculate about the way that Americans saw themselves and their future. In the longer run, these communities have had a considerable impact on the nature of American industry, lifestyle, and politics.

The chapter is based on ten years of research on primary and secondary sources, including more than a hundred interviews with defense-industry veterans, managers, scientists, and engineers. The research was designed to probe several aspects of the postwar military-industrial complex: why are firms and military facilities where they are (a locational study); how do military-oriented companies behave with respect to employment and business practices and what effect does this have on the overall performance of the economy; how were firms, workers, and communities created by the Cold War and how have they adjusted to substantial cuts in military spending after periods of war-related or other buildups? The inferences drawn about work and residence are partly speculative, and I take pains to point this out at the proper junctures.

The Character of Cold War Production

The cultural consequences of the Cold War at home are tied up with two important economic factors—changes in work and changes in residence. Europeans, whose economies are in many respects quite similar to that of the United States and were so in the 1950s, have not experienced anything like American interregional mobility and wholesale shakeup in the hierarchy of cities. Nor have the ideas of technology, meritocracy, and individual advancement so corroded the power of their labor movements. There are other determinants of these differences, many of which are rooted further back in time. But the Cold War dramatically transformed occupational and residential shifts in the United States into something qualitatively different and unique.

I begin the story by elucidating the strange quality of cold warfare, although a comprehensive account would ask the important prior question, "How did the Cold War come about?" The answer to that question would require an acknowledgment of the cultural, economic, political, and social forces that led the United States to become engaged in the pitched battle with the Soviet Union and in the nuclear arms race it entailed.[1] Although my account begins with the technical aspects of Cold War, I don't mean to argue monocausally that technology determines industry which in turn determines work and culture, but rather to write a narrative about the interaction among all of these.

I have argued elsewhere that if World Wars I and II were the first truly modern wars, then the Cold War is distinctively postmodern.[2] The differences between the two cannot be overstated. The World Wars, Korea, and Vietnam were hot wars, conducted with endless battalions of men, armed with huge quantities of guns and ammunition, and transported on thousands of land-based vehicles, ships, and fighter planes. Such "total war" was made possible by the extraordinary productivity of the modern industrial economy. Every plant in the industrial heartland complex of steel, motor vehicles, machinery, and chemicals pumped out materiel for the war effort. Innovation was often foregone in the push to turn out airplanes and equipment faster. Quantities of armaments, and the esprit and readiness of troops, determined the outcome.

In contrast, the Cold War era has been characterized by the replacement of manpower with highly sophisticated, electronics-intensive, deadly precision machinery. Machines, not men, form the core of a nation's Cold War readiness. As the expense and elaborateness escalate for each submarine, bomber, and missile, fewer and fewer units of each are produced. Each is a product of the fusion of science, technology, and engineering that have become central to Cold War military competition, where each competitor strives to stay one step ahead of its adversaries. New generations of weaponry follow fast on the footsteps of the last, as the Stealth or B-2 bomber has the B-1. Rather than responding to foreign policy priorities, strategic planning for weaponry takes place, far in advance of foreign policy formation, in the laboratories of giant corporations and major research universities, and on the drawing boards of military users. The possibility of science-based capital-intensive warfare has dramatically altered both the way the United States equips itself for war and the way it conducts diplomacy. Its "products," like the Stealth bomber, the Trident submarine, and the MX missile, along with their extraordinarily complex payloads, are the major innovative works of Cold War economy.

Why is this Cold War industrial complex so different from that which predated it (and those portions that still pump out tanks, oil, and uniforms for troops)? First of all, the shift to atmospheric warfare required a set of technologies that differed radically from those used in land- and sea-based warfare. The need to deliver airborne explosives and materiel over long distances, quickly, without being detected, and under all possible weather conditions drove the development of early radar, computing, and sophisticated communications and guidance systems that form an increasing portion of the payload in any big aerospace system. Whole new industries emerged to serve these missions; some of them went on to spin off robust commercial sectors.

Second, the push toward removing human labor from the delivery and guidance of Cold War weapons, epitomized in the ICBM missile and the communications satellite, further enhanced the role of defense electronics. Consider, for instance, the Maverick missile, a prime example of automated weaponry. The Maverick is designed to "see" targets and select which ones to hit, relying on complicated and miniaturized electro-optical equipment. To accomplish this, it carries a tiny television camera embedded in its nose, and its response system can be programmed to "kill" tanks, bridges, or other specific shapes that come into view. The Maverick incorporates the latest in electronic components, integrated circuits, and minicomputers for signal processing, carrying many thousand times more computing power and sensory capability than did its predecessors.

"Our missiles," bragged Norman Augustine, a former assistant secretary of the army and then vice president of Martin-Marietta, the company producing the Maverick, "will be not just smart, but brilliant."[3] Paradoxically, this shift toward capital-intensive warfare required a shift toward skilled labor-intensive weapons production, as research and development, design, testing, and evaluation of new guidance and communications technologies became increasingly important.

Third and finally, the Cold War, inextricably linked to nuclear weapons, set off a historically unprecedented process of relentless, government-underwritten innovation. Cold War is won when victory is achieved by besting an opponent's weaponry—its delivery system, its lethality, its detectability, its flexibility, its directability. Each breakthrough poses new problems for the opponent, who then struggles to match the virtuosity of a new-weapons technology until ultimately the initiator's competitive edge is eroded. Generation after generation of high-tech-directed nuclear weapons, launched from a battery of sophisticated delivery craft, were drafted and built, but for a strange mission—that of deterrence. The point was no longer to produce and use lethal weapons in large numbers, but to produce a few of such deadly superiority that they would ensure their own disuse. In short, they were constructed for their non-use, rather than their use value.

To match this accelerated and relentless pace of change, military leaders developed "a new awareness of invention," in the view of historian Martin van Creveld. Where once inventions were unexpected, increasingly normal preparations for war became a process of managing technology. The traditional resistance of military managers to innovation was irrevocably eliminated. In contrast, the modern military manager became centrally preoccupied with choosing paths of technological development and managing the organizational changes necessary to make the new weapons systems work.[4] Meanwhile, as Gregory Hooks has shown, the soon-to-be Pentagon appropriated the legitimacy and methods of New Deal centralized planning functions (simultaneously dismantling the social planning agencies) into a large new oversight structure.[5] Charged with strategic bombing and ballistic missile "defense," the newly independent Air Force became the ascendant service, with its culture of working with private contractors rather than producing military equipment in government arsenals.[6] As this large and unique endeavor became more institutionalized and permanent, the complexion of the military-industrial work changed dramatically.

Cold War Workers

The Cold War innovative dynamic became institutionalized in what we have come to know as the military-industrial complex. Where previously most innovations in equipment for war were adoptions of domestic or industrial technology, innovation was now to be conducted in taxpayer-funded laboratories, mostly inside private, for-profit companies, expressly devoted to military endeavors. This empire-building took place during an enormous upsurge in the American economy that permitted considerable elasticity in government spending. Where previously large commitments of public moneys had been made only in wartime, relatively high levels of the gross national product—between 5 and 10 percent annually—were now devoted to military preparedness. Between 1891 and 1916, defense spending absorbed only 1.3 percent of the gross national product on average, and from 1921 to 1940, just 1.2 percent.[7]

Owing to the nature of the arms race and the demands of atmospheric and remote-guided warfare, weapons' production during the course of the Cold War became more craftlike, the polar opposite of mass production to predrawn specifications. Considerable state-paid labor goes into pure conceptualizing, including basic science research that might contribute to useful military equipment. More labor goes into planning the process of experimentation and prototype building, and even more into testing and evaluation. Additional labor goes into security arrangements and into the back-and-forth negotiations with the military customer. To perform these functions, great numbers of scientists and engineers were accumulated in specialized firms that perform both the research and development and the ultimate production of each successive generation of arms.

As direct warmaking became more highly automated, the business of outfitting for war paradoxically became more labor-intensive, although labor of a particularly, highly skilled type. In the era of mechanical warfare, industrial output was expected to arm or transport soldier power. During the Cold War, the point became to replace it. Where factories once turned out mass-produced rifles and tanks by the thousands, the craftlike production of the handful of missiles, satellites, bombers, fighters, and nuclear-powered submarines began to take place in structures that resembled office buildings more than factories. Supporting casts of thousands are needed for each—scientists to invent the weapons, engineers to design them, machinists to build them, more engineers to test them, programmers to direct them, and technicians to repair them.[8]

Ironically the evolution of an automated Cold War threat did not lower the number of people dedicated to war readiness. Instead, it pushed them back from the uniformed front lines into roles as mechanics and repairmen, computer programmers, procurement officials, and growing phalanxes of private-sector professional workers. In the Civil War era, nine out of ten soldiers belonged to combat-ready units. In the Cold War, only one in six did, a phenomenon known in military circles as the "declining teeth to tail" ratio.[9] With Cold War, the business of killing became more remote, and the deployment of an effective threat became predominantly an industrial and scientific undertaking.

The scientist and engineer became the war hero, not the grunt or the army officer. It is not too much to wonder whether the disaffection in Vietnam with both soldier and superior did have something to do with this emerging enamoration with "clean" science- and engineering-based warfare. In this dramatic change in perceptions, the Manhattan Project played a pivotal role. Secret and lavishly expensive, the ultimate in a "trust us" gamble, the Manhattan Project employed 150,000 scientists and technicians at its peak. The success of the A-bomb investment appeared incontrovertible, despite evidence that its use was not required to win the war against Japan. After the war, the project became institutionalized in the Los Alamos National Laboratory where today it enjoys a budget of $1.2 billion a year, almost 50 percent higher than the annual budget of the Massachusetts Institute of Technology.[10]

This new contingent of scientists and engineers, wrapped in the aura of success of the Manhattan Project and later the space program, constituted the single most important new occupational group in the society. Each new weapons system required on average about $1 billion in research and development costs alone. As new weapons systems appeared on the drawing boards, funds for hiring scientists and engineers were appropriated, and universities were encouraged by the National Defense Education Act to expand educational programs in military-relevant fields like electrical engineering and computer science, especially with generous graduate student and research support. By the early 1960s, roughly half of all 1.2 million American scientists and technologists were supported by the federal government budget, either directly as civil servants or indirectly through contracts with companies, nonprofits, or universities.[11] This was a radical departure from previous decades, when public funding accounted for less than 5 percent of research and development in the United States.[12]

The complexion of this new science and engineering workforce was quite distinct from that serving commercial industry. Up through World War II, engineering occupations were clustered in three, perhaps four, areas: civil engineering, devoted to the building of infrastructure; mechanical and industrial engineering, devoted to transportation and mass production processes; chemical engineering, devoted to substances from drugs to plastics; and electrical engineering, devoted to electric power and communications. With the Cold War, electrical engineering was transformed into electronics engineering, and previously small occupations like aeronautical engineer and oceanographer were expanded exponentially. The ranks of scientists in fields like nuclear physics and astronomy began rapidly to outpace those in other less military-relevant fields.

By the 1980s this increment had created a science and engineering establishment where the more esteemed and well-paid occupations were those linked to Cold War innovation (Table 1).[13] These were the fields with the most glamorous, exciting work. Take oceanographers, for instance, the profession so appealingly profiled in Jacques Cousteau's television work. In reality, this occupation was closely associated with submarine warfare and the need to study the conditions under which submarines could serve as the most invulnerable of the nuclear triad—as the mobile and undetectable launching pads for ICBM missiles. While aero- and astro-nautical engineers are the most defense-dependent, a surprising number of physicists, astronomers, mathematicians, and other physical scientists are also devoted to Cold War innovation.

The Pentagon and its defense contractors were willing and able to pay a premium for science and engineering labor. In the mid-1980s, at the peak of the Reagan buildup, defense-supported scientists and engineers were paid 17 percent more than those working in nondefense related jobs, a gap that doubled in the 1980s. These wage differentials consistently drew talent away from the engineering and science occupations that are devoted to commercial vigor in the economy, a depletionary process analyzed at length by Seymour Melman and Lloyd Dumas.[14]

In part paying premiums has been a necessity, since young scientists and engineers have demonstrated some reticence to signing up with defense contractors

Table 1.
Defense Dependency by Occupation, 1986

Occupation	Percent Defense-Dependent
Aeronautical and Astronautical Engineers	69
Oceanographers	50
Physicists, Astronomers	34
Electrical and Electronics Engineers	32
Metallurgical and Materials Engineers	24
Mathematicians	20
Physical Scientists	20
Mechanical Engineers	18
Other Engineers	17
Nuclear Engineers	16
Industrial Engineers	13
Computer Scientists	13
Atmospheric Scientists, Meteorologists	10
Statisticians	9
Civil Engineers	8
Chemists	7
Earth Scientists	5
Chemical Engineers	5
Other Social Scientists	5
Medical Scientists	4
Economists	4
Biologists	4
Sociologists, Anthropologists	4
Biochemists	3
Psychologists	3
Mining Engineers	3
Petroleum Engineers	1
Agricultural Scientists	1

Source: Compiled by Ann Markusen from unpublished data from the National Science Foundation's Survey of Scientists and Engineers and the Bureau of Labor Statistics.

Note: For an explanation of the method of estimation, see Ann Markusen and Joel Yudken, "The Labor Economics of Conversion: Prospects for Military-Dependent Engineers and Scientists," in Patricia MacCorquodale, Martha Gilliland, Jeffrey Kash, and Andrew Jameton, eds., *Engineers and Economic Conversion* (New York: Springer-Verlag, 1993), 135–91.

and the armed forces for both political and career reasons. There has been a critical and vocal contingent within the senior ranks of the science and engineering professions that has worked assiduously to control the arms race and to limit domination of academic science and engineering by the Pentagon, beginning with the atomic scientists' rebellion against the H-bomb in the 1950s and demonstrated most recently in the controversy over Star Wars.

The impact of Cold War on occupations has not been confined to boosting those in certain branches of science and engineering. It has simultaneously accelerated the displacement of blue-collar workers in commercial manufacturing through the application of its tools of automation to the factory floor. Automated, remote controlled warfare had its analog in associated spin-offs in CAD/CAM systems, robotics, and numerically controlled machine tools. These were developed much more rapidly than would have been the case without defense-bred innovation. Over the postwar decades, they spread like wildfire through the civilian production sectors, displacing workers with computer-driven programs and increasing productivity so rapidly that the size of the manufacturing workforce declined markedly even though absolute levels of output remained steady.[15]

Even in the defense-related sectors like aerospace, the proportion—the absolute level—of blue-collar workers has declined over the decades as design intensity, automation, and more expensive weapons produced in smaller lots have become more prominent. Although aerospace employment—more than 60 percent defense dependent—grew steadily through 1987, there were 40 percent fewer production jobs in 1987 than in 1968, a net loss of 440,000. Production workers as a percentage of total workforce fell from 53 to 45 percent.[16]

Exacerbating the economywide loss of blue-collar jobs was the relative competitive disadvantage thrust on nondefense industries in the United States through the bidding up of resources by defense contractors. Firms in steel, auto, machinery, and electronics had to compete for capital and labor with defense contractors, and had no comparable policy blend of research and development, procurement, and defense-industrial base stabilization devoted to their industry. As European machine tools and Japanese consumer electronics and autos outcompeted American producers, jobs in these blue-collar industries declined dramatically, especially during the 1980s buildup.[17]

Blue-collar displacement plus the swelling of the scientific and technical workforce in defense-oriented sectors may be important contributors to the marked worsening of income distribution in the United States. Since the early 1970s, the relatively "good" American income distribution has deteriorated with the disappearance of what Bennett Harrison and Barry Bluestone call the "missing middle," that segment of well-paid, often unionized blue-collar workers who made enough to support a family and consider themselves "middle class."[18] This is ironic, since top union leadership strongly supported the Cold War throughout most of the Cold War period.

This Cold War occupational restructuring has left its mark on the social and psychological profile of the workforce as well. The rise in the Cold War science and engineering professions brought a swelling in the ranks of jobs that is heavily skewed toward educated white males. The engineering professions are more heavily segregated by gender and race than almost any other occupation in the United States. African Americans made up only 3 percent of the science and engineering workforce in 1988, and women only 15 percent, far below their participation rates in the workforce overall. Much of this is explained by discrimination in educational and hiring practices in science and engineering generally.[19]

Among defense-dependent scientists and engineers, the degree of representation for women is even worse, and wage gaps run even wider. In the heavily defense-dependent Los Angeles aerospace industry, full-time working women make only 51 percent of what men do, compared with 64 percent in the economy as a whole. For minorities, however, if one manages to surpass the educational barrier, one's chances of landing a defense-dependent job are actually slightly greater than in the nondefense sector, presumably because the government is more rigorous in its affirmative action practices.[20]

For blue-collar workers, discrimination and long-term shrinkage in the ranks have led to defense plant workforces dominated by older white males. Minorities and women were heavily recruited to work in naval shipyards and aircraft plants in World War II, but after the war, jobs were disproportionately retained by returning veterans and male workers. Meanwhile, the dramatic decline in blue-collar jobs in nondefense industries, where minorities and women had made substantial inroads in the 1980s, has again cut into their ranks. For urban minorities in plants in the industrial heartland, stretching from Buffalo through Milwaukee, the displacement has been particularly fierce, eliminating jobs for younger generations of youth who have involuntarily joined the ranks of the "underclass."

Psychologically, the juxtaposition of the new Cold War occupational increment with the related corrosion in the ranks of blue-collar workers altered fundamental attitudes toward work in this country. Cleverness, competitiveness, individual achievement, and locational mobility, as we shall see, came to be valued above the older industrial values of hard work, consistency, solidarity, and respect for seniority associated with blue-collar work, values forged over decades of negotiated conflict between management and organized labor. The dignity of manual labor was depreciated in favor of the braininess and brilliance in a new form of meritocracy, which in turn bore some of the culture of military hierarchy with it (especially since the personnel systems of defense contractors tended to be headed by former military top brass). It is impossible to gauge the extent of this Cold War–induced change in attitude toward work, which has other roots as well. It is enough to say that in the United States, blue-collar work today is held in much lower esteem than in any other advanced industrial country.

Cold War Communities

As the Cold War waxed, a relatively unique set of communities thrived on the specialized activities associated with automated, atmospheric warfare and the arms race associated with it. For strategic and other reasons, new sites outside the traditional industrial heartland were favored by military officers and contractors in charge of locating and operating defense-related facilities. In an arc stretching from Seattle down through California, the Intermountain West, Texas, isolated spots in the Southeast, and up the eastern seaboard to Long Island and New England, the cities and regions form what we have called elsewhere "the gunbelt," accounting for a major increment in population and jobs in the postwar period. Among them are localities that captured the American imagination as either sunny, suburban paradises (Los Angeles, Orange County, San Diego) or yuppie, high-tech enclaves of the future (Silicon Valley, Route 128 outside Boston, Seattle). Greater Los Angeles, home also to Hollywood and Disneyland, became the military-industrial capital and center of gravity of the gunbelt.

Several forces were at work in shaping the gunbelt.[21] First, military bases, especially those associated with the Air Force, had traditionally been sited in relatively remote areas where land was cheap and secrecy could be assured. New military activities—both public and private—gravitated toward these sites, which were disproportionately in the South and the West.

Second, strategic considerations in the decades of the 1940s and 1950s helped to develop the West Coast aircraft industry, oriented toward the Pacific front and complementing the strong naval presence that had been there since the turn of the century. Strategic concerns—fear of bombing in particular—also forced a decentralization of activity away from coastal regions and into the interior. St. Louis, Wichita, Los Alamos, Colorado Springs, and other cities benefited from the military's concern that defense plants and key laboratories and command centers be well-protected after the Pearl Harbor experience.

Third, military officers and entrepreneurs in the new aircraft and electronics industries were insulated from the type of locational cost pressures that kept most other industry agglomerated in the industrial heartland. Top Air Force brass, in particular, exhibited a strong repulsion from East Coast cosmopolitan society and preferred the wide open spaces and frontier mentality of the West, demonstrated in their choice of Colorado Springs for their new academy. They were able, through discretionary power over the location process, to influence the geographical distribution of military-related activity.

Fourth, civic leaders and local business interests in the South and the West were active boosters in enticing new aircraft and related facilities, including military installations, to their environs. The story of the competition among San Diego, Los Angeles, and San Francisco for naval facilities has been masterfully told by Roger Lotchin.[22] In Colorado Springs and other western cities, and in the South, facilitated by congressional longevity, boosterist coalitions were able to win sizable commitments of government contracts that permanently brought them infrastructure and economic activity.

The outcome of this locational process, which began between the two World Wars but was solidified in the 1950s, is apparent in the net shifts in defense contract awards in the postwar period (Table 2). In 1952, during the Korean War, prime contracts were most heavily concentrated in the old manufacturing belt—New England, the Middle Atlantic, and the east north-central states, or Boston through Chicago, plus the Pacific states. All other regions fell more than 20 percent below the national average in per capita prime contract receipts. In the 1950s, however, the shares of the Middle Atlantic and east north-central begin to fall dramatically, while the Pacific region and all other outlying regions increase their shares substantially. New England, hard pressed by deindustrialization in textiles, shoes, and machining, converted itself into the premier defense-dependent region by diversifying into everything from aircraft engines to radar and guidance systems to nuclear weapons and naval ships.

Within these regions, certain cities were the major beneficiaries of military-related growth (Table 3). Three of the top four counties are in California—Los Angeles, Santa Clara (Silicon Valley), and Orange County. Suburban Boston and Long Island each have two counties in the top twenty, and Connecticut, three. San Diego, Seattle, Dallas, Jackson (Mississippi), Atlanta, and Wichita make the list, along with a few older cities—St. Louis, New York, Baltimore. Smaller cities, too, thrived on the Cold War effort. Some, like Santa Fe (Los Alamos), Oak Ridge, Livermore, Hanford (Washington), Houston, or Titusville/Melbourne (Washington) do not show up on this list only because their Cold War funding comes from the Department of Energy or the National Aeronautics and Space Administration. Other smaller cities, like Colorado Springs, Huntsville (Alabama), Bath (Maine), Charleston (South Carolina), and Newport News (Vir-

Table 2.
Department of Defense Prime Contract Awards* Per Capita Spending Relative to U.S. Average by Region: Fiscal year 1952–1986**
(Index numbers, U.S.=100)

Region	1952	1958	1962	1967	1972	1977	1982	1985
U.S. Total	100	100	100	100	100	100	100	100
New England	136	163	175	195	165	216	219	214
Middle Atlantic	125	101	101	92	103	88	82	82
East North Central	133	67	62	70	55	47	48	52
West North Central	79	92	75	102	100	115	110	107
South Atlantic	53	68	74	83	90	80	94	93
East South Central	23	26	30	51	67	71	54	54
West South Central	49	85	65	120	106	84	97	90
Mountain	20	87	135	60	77	68	68	75
Pacific	179	235	228	157	177	193	177	168

Sources: "Prime Contract Awards by Region and State," annual, 1951–1986, Department of Defense; "Current Population Reports," series P-25, selected years, U.S. Bureau of the Census.

Notes: *Prime contracts >$10,000. For fiscal years 1982–1986, prime contracts >$25,000.

**Years shown are 3-year averages, e.g., 1952 equals average 1951–1953.

ginia) also have become dominated by Cold War activities, as are dozens of rural counties hosting Air Force bases, missile silos, and nuclear testing facilities.

This marked geographical shift is closely associated with the great divide between hot war and cold war outfitting.[23] During hot wars, ordnance, tanks and trucks, guns and ammunition bound for blue-collar soldiers were pumped out of industrial belt factories by blue-collar workers. The same companies that made automobiles or cameras—General Motors, Ford, Bell, and Howell—converted themselves into military-serving plants during hot wars, using the same assembly-line techniques to make everything from bombers to optical equipment. At war's end, especially after World War II, they quickly and agreeably reoriented their plants and workforce back to consumer and commercial goods production, glad to be rid of cumbersome and uncertain negotiations with a single government customer. Pent-up demand and orders associated with the rebuilding of Europe, Japan, and Korea helped. Only New England, which had lost its manufacturing leadership to cities west of the Appalachians, found it necessary to pursue Cold War outfitting as a revitalization strategy.

The western aircraft industry with its youthful suppliers in the guidance, electronics, and computing industries was in an altogether different position. From

Table 3.
Top-ranking Counties, Military Prime Contracts, 1984

Contracts		Total
Rank	County	Billions of Dollars
1	Los Angeles, California	13.8
2	St. Louis, Missouri	5.9
3	Santa Clara, California	4.7
4	Orange, California	3.7
5	Nassau, New York	3.6
6	Middlesex, Massachusetts	3.6
7	Tarrant, Texas	2.9
8	King, Washington	2.4
9	San Diego, California	2.2
10	Cobb, Georgia	2.1
11	Sedgwick, Kansas	2.1
12	Essex, Massachusetts	1.9
13	Hartford, Connecticut	1.8
14	Jackson, Mississippi	1.8
15	New York, New York	1.7
16	New London, Connecticut	1.6
17	Fairfield, Connecticut	1.5
18	Suffolk, New York	1.4
19	Dallas, Texas	1.4
20	Baltimore, Maryland	1.4

Source: U.S. Department of Commerce, Bureau of the Census, 1984.

its inception the industry had been heavily government-dependent: commercial airlines were tiny and traffic too small to cover the extraordinary design and development costs. The industry had been subsidized since World War I by government research and development and procurement contracts.[24]

As World War II approached, orders rose dramatically with Roosevelt's pledge in 1940 to construct 50,000 airplanes a year. Aircraft plants expanded with incredible rapidity, producing 96,000 airplanes in 1944. Employment in the North American (now Rockwell) plant in Los Angeles, for instance, zoomed from 6,000 in 1940 to 92,000 in 1943. After the war, the process abruptly went into reverse. Over 90 percent of contracts were canceled, and employment fell from over 1 million to a low of 238,000 in 1948.[25] Aircraft companies were desperate. Most of their facilities were isolated from centers of robust commercial activity in the industrial heartland. Conversion was not an option, they felt. They organized to press vigorously for public commitment to air defense systems and government assurance of an industrial base in aircraft.[26] Quietly, with their connections in the Air Force and organizations like the Rand Corporation, they helped to build the rationale for the Cold War.

The American industrial economy thus traveled along two very different routes in the 1950s. In the industrial heartland, mass production, commercial corporatism, and working-class cultures dominated the immediate postwar period, dismissive of the gunbelt empire-building that was to undermine their future viability. In the gunbelt, new prototypically suburban communities evolved to house the extensive research and development, coordination, lobbying, and design activities associated with the new "postmodern" form of warfare. Just how were these communities formed, and how did they differ from the rest of the society?

The archetype of the Cold War community was Los Alamos. Hurriedly thrown together during the early 1940s at a boys' boarding school as a highly secretive site for the development of the atomic bomb, Los Alamos defied most criteria that guide the city-building process. It was extraordinarily inaccessible, without the transportation corridors—railroads, highways—that generally spawn towns. No preexisting commercial sector was there. It was intended to be a temporary facility only, and the buildings and housing constructed were inexpensive and barrackslike.[27] Fifty years after its birth, Los Alamos still bears some of the features of a temporary community, much of the growth it spawned displaced to nearby Santa Fe and Albuquerque. Yet Los Alamos turned out not to be temporary, but a huge and crucial facility in the nuclear weapons development process.

The wrenching of scientists and engineers away from cosmopolitan society for a project of this sort served as an experiment in a new form of professional life and community. In interviews we have done with military-related managers and scientists in a number of locales in the West, many expressed an emphatic preference for wide open spaces and enclave communities reminiscent of Los Alamos.[28]

A number of cities assiduously worked to land Cold War defense facilities. Boosters in Colorado Springs, for instance, cultivated the military top brass by providing "rest and recreation" for them during World War II and parlaying this into a winning bid for the Air Force Academy, sweetened by the offer of an extensive tract of free land. They successfully fought to be the home of the Air Defense Command, eventually a key facility in the Star Wars effort. These facilities, plus a large combat-ready army base, Fort Carson, acted as anchors that subsequently pulled in a number of key defense contractors (Kaman

Sciences, Ford Aerospace) as well as branch plants of crowded California electronics firms.

A key drawing card for a city like Colorado Springs was its attempt to be military friendly. As early as the late 1930s, when the town was losing its economic base in mining and agriculture, the Chamber of Commerce formed a Military Affairs Committee whose job it was to go after some of the loose cash the Army Air Corps was spending on new airfields around the country. During the Air Force Academy recruitment drive in the early 1950s, the chamber coached every taxi driver in town to respond favorably to queries about public attitudes toward the military. The Military Affairs Committee has continued to work over the decades, anticipating potential problems between military tenants and the community and continuing to lobby for additional military commitments.[29] Dozens of other communities throughout the South and West have similar organizations dedicated to enhancing relationships between the military and the local population.

The defense-dependent metropolis is a companion to the isolated military-industrial city. Despite its frontier image, the West is actually more urbanized than any other region, and it was in a number of these cities—Seattle and Palo Alto to San Jose, Los Angeles/Orange County, San Diego, San Antonio, and Dallas—that the most important increment of new Cold War–related activity emerged.[30] Large Cold War plants and offices were relatively suburbanized from the start, following in the wake of the youthful aircraft industry.[31] The prototypical community created by this Cold War activity was Orange County, the dream environment of the 1950s, a land of affluent white people, barbecues, Bermuda shorts, oranges, and surfing. The role of defense contractors in creating Orange County is curiously underplayed in the literature, which depicts it as a "post-suburban," information-based community.[32] Looking for more space and the kind of environments that scientists, engineers, and managers were supposed to prefer, companies like Northrop, Hughes, Lockheed, and Ford Aerospace moved or started up huge operations in Orange County in the 1950s, bringing with them a large increment of upscale and reliable jobs.

A striking aspect of this process was the way in which the Cold War labor force was created, relying heavily upon interregional migration of highly educated people to new Cold War enclaves. World War II had started this process, as workers of all sorts were recruited to western and southern defense facilities. Engineers were disproportionately represented among the movers. A study in the early 1950s found that 59 percent of engineering majors had left their home states over the previous decade, almost twice the rate of nonengineering majors, which was again higher than noncollege graduates, and that net migration flowed heavily from the East and

Midwest toward the West and the South.[33] By the end of the 1950s, the top engineering schools of the Northeast/Midwest, most of them publicly funded, were exporting more than half of their engineering graduates—the most expensive to educate—out of state. Work we have recently done on the Census and Survey of Scientists and Engineers data base show that this internal brain drain continued over the decades, creating large, new high-tech labor pools in selected southern and western cities. Defense-related scientists and engineers moved farther and to different locations than did those who were not defense-dependent.[34]

This selective process of migration was underwritten by American taxpayers. In the 1980s it cost between $5000 and $20,000 to move a defense-related engineer from one locale to another. Over the decades, tens of thousands of high-tech personnel had been recruited with such relocation assistance. This cost was subsumed in defense contractors' cost structure, part of what the government paid for in "cost-plus" defense contracting. Ironically, defense contractors thus made a profit on the relocation of skilled personnel. Although this was not the intent of Pentagon practices, the interregional movement of scientists and engineers was in effect the largest, publicly underwritten, for-profit population redistribution scheme in American history.

Although this account remains incomplete, the Cold War was the seminal event in gunbelt community-building. The 1950s, as we have shown elsewhere, was the key decade in which the gunbelt took shape, setting up the duality between the older, commercially oriented and deindustrializing heartland and the new, high-tech, government-dependent periphery. Individual communities were highly colored by their particular expertise and role in the Cold War. They shared a technocratic culture, an acceptance of Cold War ideology, a relatively conservative politics, and a pollyannish view of the future of United States society, a view that proved to be unwarranted.

Cold War Occupations, Cities, and Culture

There may be strong connections between these new occupations and enclaves and the shaping of postwar American optimism, conservatism, militarism, and consumerist political culture. I draw a number of conclusions about the consequences of new Cold War work and residential patterns. They remain speculative, more in the form of arguments, and are offered in the spirit of encouraging further research.

Although the affluence of the new Cold War–driven communities—and the related deterioration of the industrial heartland—contributed to a superficial improvement in interregional income distribution, the process of Cold War restructuring has had other negative consequences. First, the rise of the gunbelt has been costly to the nation in economic terms, exacerbating economic performance problems, draining the federal treasury, and quickening the urban crisis in older industrial centers.[35] Second, the gunbelt dramatically changed American geopolitics, creating new, relatively conservative political regions that account for an increasing share of the electorate and are increasingly in a position to guarantee the maintenance of militarism. Third, the gunbelt may have contributed to an unwarranted optimism about the nation's future and to the development of a consumerist, individualistic, apolitical outlook that supplanted the New Deal emphasis on the dignity of work, community, activism, and social security.

First, the economic costs. The gunbelt consists of a series of highly specialized regional enclaves devoted to Cold War armaments. Because these enclaves tend to be physically far removed from commercially oriented industries elsewhere, both types of host communities are subject to more dramatic swings in local economic activity than would occur if they were more integrated and diversified across defense and nondefense activities. Political cycles of defense buildup and contraction do not correspond very closely to business cycles as we know them. If defense-oriented industries were located in tandem with commercial ones, then we might expect greater regional resiliency to unemployment generated from downturns in either sector.

Historically, when the major military supply industries were located in the industrial heartland, this convertibility worked rather well. The enormous diversion of resources into and out of defense production before and after World War II was handled quite easily, as Detroit retooled auto plants into bomber factories and Illinois retooled tractor plants into tank factories. But as defense production shifted away from the heartland, this flexibility was increasingly lost to regions at both ends.

This is underlined by recent experience. In the past decade, the United States has experienced a remarkable divergence in regional growth rates. From 1979 to 1986, when defense procurement rose rapidly in real terms, jobs in New England, the South Atlantic, Mountain, and Pacific regions grew in excess of 15 percent, while in the Great Lakes states they declined by some 65,000.[36] There, in the early 1980s, a policy-induced recession was displacing workers from auto and steel plants. But no recession-proof aerospace or shipbuilding jobs were nearby to take up the slack. Throughout the postwar period, the rise of the gunbelt has meant that regional economies are more apt to be plunged into idiosyncratic economic cycles that do not necessarily reflect national trends. This is exacerbated to the extent that regions specialize in certain technologies or weapons systems.

The investment of billions of tax dollars in new, peripheral industrial centers has added to the problems faced by industries in the older, commercially oriented regions. These problems are of two kinds: a lack of basic investment in regional infrastructure, and a loss of vital creative talent. Because the new aerospace-oriented communities were erected far from the industrial heartland, considerable tax resources have gone into their construction and maintenance. While older plants in the Midwest have been shuttered and mothballed, new plants have been built in the South and the West. Around them new cities and suburbs have sprouted, often with considerable taxpayer support in federal moneys for sewer, water, highways, and other types of infrastructure. In many cases the Pentagon pays "impact aid" to school districts and local governments, compensating for the tax base that is lost through government tax-exempt property.

Billions have been spent over the postwar period in such community building. Such a social investment would not pose a problem if it were true that infrastructure elsewhere was fully employed. But over the same period, older industrial regions have experienced industrial decline and dramatic outmigration, emptying plants, houses, and school districts and lowering infrastructure utilization rates to the point where receipts make it difficult to maintain or modernize them. To the extent the new schools, plants, and utilities are paid for by taxpayers nationally, closure or underusage of similar capacity in other parts of the country constitutes a net waste of societal resources. Funds spent could have been used to rebuild infrastructure in existing cities, to better house the lower-income population, or to cut taxes, lower deficits, or improve the quality of life in many ways. If defense-related production had been more evenly distributed geographically, the nation could have saved substantial amounts of money.[37]

Already disadvantaged in the competition for government research/development and procurement dollars, the older industrial regions are further disabled by the geographical distance from these newer, more glamorous industrial complexes. The removal of large numbers of the most highly skilled and creative people from their regional economies—people who were educated with corporate and personal tax dollars from within the region—handicaps their ability to compete in increasingly competitive international markets. It results in the absence of cross-fertilizing ideas flowing through the regional business community. True, big firms in industries like auto, steel, consumer electronics, and machine tools are in large part responsible for their own disappointing performance—particularly through their oligopolistic preoccupation with dominating their markets. Nevertheless, their workers and host communities are doubly debilitated by the absence of higher-tech alternatives nurtured in the new defense-industrial complexes.

This mass migration of talent to the gunbelt appears also to have contributed to the increasing spatial segregation of America by class and race, for it has been highly selective. Defense-related industries, as we have seen, are more apt to employ white, highly educated men than are other manufacturing industries. Although during World War II large numbers of blue-collar workers were also relocated, the need to do so has disappeared over time, leaving only the moves of more affluent employees to be taxpayer-financed. Furthermore, the shares of blue-collar workers in the defense-related workforce have dropped dramatically. Communities that have been built around defense plants and offices disproportionately consist of members of the white-collar, professional/technical social stratum and their families: the gunbelt yuppies.

The other side of the coin has been the creation of the urban "underclass." This is a relatively recent phenomenon. It consists principally of labor pools of poor blacks and whites from the South and Appalachia who came to northern cities (and to California) as recently as the 1950s in search of good industrial jobs which were then plentiful. But by the end of the 1950s, the momentum in favor of peripheral gunbelt sites gathered speed and manufacturing jobs permanently shifted their centers of gravity. Industrial plants in civilian markets closed, leaving behind dislocated workers and their children, often in inner-city ghettos.

My second argument concerns the geopolitical consequences of the rise of the gunbelt. The shift of the military-industrial complex to the gunbelt has been accompanied by a simultaneous shift of political power toward regions that are markedly more conservative and Republican than the industrial heartland.[38] Over the postwar period, the center of gravity of American politics has shifted farther and farther South and West, paralleling the movement of leading-edge manufacturing capacity. Decade after decade, redistricting gives greater power in the Electoral College and in the Congress to voters from newer defense-dependent districts. By and large, white-collar employees of defense industries vote much more strongly Republican than do their counterparts in the rest of the economy. It is no coincidence that southern California is a safe constituency for promilitary conservatism and the home of the John Birch Society, or that Colorado Springs consistently elects hawks and votes 70 percent Republican. The emergence of new military-industrial centers in the Old South has contributed to the recomposition of the southern electorate and its shift toward Republican presidential candidates, as they pulled in more highly educated white professionals while poor, rural, black Democrats left for the North.

Nowhere is this more apparent than in the presidential vote. The states of California, Texas, and Massachusetts account for no less than five of the nine pre-Clinton postwar presidents. Each state has reaped a disproportionate share of the nation's defense contracts. All of the presidents elected in the postwar period hail from states outside the industrial heartland. Only Gerald Ford, who was never elected, represented a heavily industrialized heartland state. Only Ford, Carter, and Clinton hail from states whose per capita shares of prime contracts were less than the nation's, although Georgia ranks exceptionally high in military facilities. In presidential races, the defense-dependent southern and western electorate in particular show greater preference for Republican candidates who promise a strong defense and are staunchly committed to major weapons systems.

The spatial segregation of defense-oriented aerospace from other industrial regions establishes an economic dualism which makes it difficult to distribute defense contracts more evenly. Members of the U.S. Congress from nondefense regions are less interested in defense matters and tend to shun committees which make the major decisions, preferring instead to serve on committees like agriculture, interior, labor, or social services.[39] They surrender the playing field to their more concerned colleagues, who specialize in the intricacies of defense spending and are thus in a better position to defend high levels and to funnel them to their home districts. The rise of the gunbelt has created its own self-sustaining mechanism. By regionally differentiating participation in the procurement decision-making process, it has reinforced and expanded the imbalances that first emerged in the 1950s.

There are even wider political implications. Because gunbelt cities have been such successful ensembles of creative talent, although underwritten almost entirely by the Pentagon, they have generated ever more stunning innovations in the power, accuracy, and speed of strategic weapons and their delivery systems. Science, not foreign policy, has come to drive weapons development. Each new generation of weaponry exacerbates the threat and sends the armed forces scurrying to secure appropriations for yet further research for arms to defend against these same systems and to develop the next offensive one. The clusters themselves, then, are part of the reproductive apparatus of strategic warfare. Even in the 2000s, after the Soviet threat has evaporated, new threats are envisioned by southern California's Rand Corporation and conservative "military preparedness" groups.

The segregation of defense-based activities from commercial industrial centers and the concentration of defense-dependent voters in new regions may thus have had a strong positive feedback effect on the level and geographic distribution of defense expenditure. Members of Congress from heavily defense-dependent districts tend to work their ways onto key military appropriations committees. Through the logrolling process, they often maximize the value of their votes by trading them for the support of others for key military projects.

Politicians who represent defense-hungry local economies can be expected to vote for appropriations on the basis of their employment and profit-generating potential, rather than on strictly strategic grounds.

Such actions are likely to lead to increases in defense budgets beyond the level that would be set by foreign policy and national political considerations alone. Overall the gunbelt may operate as a mechanism that exacerbates the pressures for military spending and the continuation of Cold War foreign policy beyond what would exist if these industries were in the same places as those that make food, clothing, producer machinery, and consumer goods, where their political clout would be diluted. We may, like Frankenstein, have fashioned places which increasingly control us rather than vice-versa.

My third argument, and it would better be termed a hunch, is that Cold War economic activity, through its promotion of certain occupations and regions at the expense of others, helped to dismantle a New Deal culture preoccupied with concern for workers, community, activism, and social security and put in its place the consumerist, individualistic, apolitical culture we have come to associate with the 1950s. Popular culture, purveyed through a Hollywood conveniently situated in the heart of the new Cold War military-industrial complex—and cleansed so effectively through blacklisting of New Deal radicals and progressives—celebrated the new meritocratic occupations and the technologies they brought to bear on societal problems and American lifestyle. The astronauts, for instance, were a particularly glamorous expression of the new type of military-industrial worker: exceptional human beings, highly educated, competent with cybernetic machinery, selected purely on the basis of their personal accomplishments and then embedded in a complex, hierarchical institution. Movies and television also honed in on the new southern Californian suburban culture with its emphasis on the comic problems of material life, coping with changing sexual mores and parenting in the affluent society. These connections ought to be further researched in interdisciplinary research on Cold War culture.

Work by other social scientists, historians, and humanists in the past couple of decades has helped a great deal to counteract the economic determinism so problematic in both Marxism and neoclassical economics. Rather than seeing material life and economic motives as the basis for all other aspects of culture and politics—what Marxists used to call, residually, the superstructure—contemporary social analysts have evolved a more even-handed approach that encourages the blending of various analytical tools across the disciplines. In this chapter I have attempted to show how the Cold War, with its peculiar demands on the structure of technology, work, and regions, has contributed to the shaping of postwar

American life and culture. The restructuring of the economy has been substantially influenced by the political, cultural, and ideological constructs of the times, as well as by other autonomous forces—social security, for instance, with its contribution to gunbelt migration. A better understanding of the distortions that the Cold War has caused in American economic and social life should help guide the nation in the coming period of post–Cold War reconstruction.

Notes

1. A good introduction to competing views can be found in Ronald Powaski, *March to Armageddon: The United States and the Nuclear Arms Race, 1939 to the Present* (New York: Oxford University Press, 1987), especially chaps. 1–6, although Powaski does not include much on the role of culture.
2. The following section draws upon chap. 2, "The Rise of Postmodern Warfare," of Ann Markusen and Joel Yudken, *Dismantling the Cold War Economy* (New York: Basic Books, 1992).
3. Cited in James Fallows, *National Defense* (New York: Random House, 1981), 56–57.
4. Martin van Creveld, *Technology and War* (New York: Free Press, 1989), 218; see also Morris Janowitz, *The Professional Soldier: A Social and Political Portrait* (Glencoe, Ill.: Free Press, 1960), 22–31.
5. Gregory Hooks, *Creating the Military-Industrial Complex* (Champaign, Ill.: University of Illinois Press, 1992).
6. Matthew Evangelista, *Innovation and the Arms Race: How the United States and the Soviet Union Develop New Military Technologies* (Ithaca, N.Y.: Cornell University Press, 1989), 56–60.
7. Michael Oden, *Military Spending, Military Power, and United States Postwar Economic Performance* (Ph.D. dissertation, Department of Economics, The New School for Social Research, 1992).
8. Mary Kaldor, *The Baroque Arsenal* (New York: Hill and Wang, 1981), 12, 25; Van Creveld, *Technology and War,* 230–31.
9. William White, *U.S. Tactical Air Power: Missions, Forces, and Costs* (Washington, D.C.: Brookings Institution, 1974), 5; Kaldor, *Baroque Arsenal,* 11
10. Powaski, *March to Armageddon,* 6; Seymour Melman, talk given at Los Alamos National Labs, August, 1993.
11. Evangelista, *Innovation and the Arms Race,* 54; H. Nieburg, *In the Name of Science* (Chicago: Quadrangle Books, 1966), 20, 122.
12. Markusen and Yudken, *Dismantling the Cold War Economy,* 102–8; Warren Davis, "The Pentagon and the Scientists," in John Tirman, ed., *The Militarization of High Technology* (Cambridge, Mass.: Ballinger, 1984), 153.
13. For a fuller discussion of the data and the formation of this workforce, see Ann Markusen, "Structural Barriers to Converting the U.S. Economy," in Jurgen Brauer and Manas Chatterji, eds., *Economic Issues of Disarmament* (London: Macmillan

Press, 1993), and Joel Yudken and Ann Markusen, "The Labor Economic of Conversion: Prospects for Military-Dependent Engineers and Scientists," in Martha Gilliland and Patricia MacCorquodale, eds., *Engineers and Economic Conversion* (Boulder, Col.: Westview Press, 1994).

14. Seymour Melman, *The Permanent War Economy: American Capitalism in Decline* (New York: Simon & Schuster, 1974), and *Profits without Production* (New York: Alfred A. Knopf, 1983); Lloyd Dumas, *The Overburdened Economy* (Berkeley: University of California Press, 1986).
15. This argument is developed more fully in Ann Markusen, "The Military-Industrial Divide: Cold War Transformation of the Economy and the Rise of New Industrial Complexes," *Society and Space* (9:4, 1991), 391–416, and in "Defense Spending: A Successful Industrial Policy?" *International Journal of Urban and Regional Research* (10:1, 1986), 105–22.
16. United Auto Workers Union, *U.S. Aerospace Industry Review* (Detroit: UAW Research Department, 1988), 8.
17. Markusen and Yudken, *Dismantling the Cold War Economy,* chap. 6.
18. Bennett Harrison and Barry Bluestone, *The Great U-Turn: Corporate Restructuring and the Polarizing of America* (New York: Basic Books, 1988).
19. Betty Vetter and Eleanor Babco, *Professional Women and Minorities*, 5th ed. (Washington, D.C.: Scientific Manpower Commission, August 1984), 37; *Science and Engineering Overview* (NSF 85-302, Washington, D.C.: National Science Foundation, 1984), 53–54; Shirley Malcolm, *Women in Science and Engineering: An Overview* (prepared for the National Academy of Sciences, Washington, D.C.: American Association for the Advancement of Science, September 1983), 27, 37; Lilli Hornig, "Women in Science and Engineering: Why So Few?" *Technology Review* (November–December, 1984), 40. See the summary of this work and others in Office of Technology Assessment, *Demographic Trends and the Scientific and Engineering Work Force* (Washington, D.C.: Government Printing Office, December 1985), 114–21.
20. Markusen and Yudken, *Dismantling the Cold War Economy,* 163–65; Ann Markusen and Scott Campbell, "A Separate Military-Industrial Job Market? Scientists and Engineers in Military-Related versus Civilian Sectors in the 1980s" (Working Paper, Project on Regional and Industrial Economics, Rutgers University, March 1994).
21. These points are made at much greater length in Ann Markusen, Peter Hall, Scott Campbell, and Sabina Deitrick, *The Rise of the Gunbelt* (New York: Oxford University Press, 1991). The difference in locational orientation when government is the market is developed in Ann Markusen, "Government as Market: Industrial Location in the U.S. Defense Industry," in Henry Herzog and Alan Schlottmann, *Industry Location and Public Policy* (Knoxville: University of Tennessee Press, 1991), 137–68.
22. Roger Lotchin, *The Martial Metropoles: U.S. Cities in War and Peace* (New York: Praeger, 1984).
23. For a fuller account of this military-industrial divide, see Markusen, *Government as Market,* 1991.

24. Howard Mingos, "The Rise of the Aircraft Industry," in G. R. Simonson, ed., *The History of the American Aircraft Industry: An Anthology* (Cambridge, Mass.: M.I.T. Press, 1968); and Markusen and Yudken, *Dismantling the Cold War Economy,* chap. 3.
25. Benjamin S. Kelsey, *The Dragon's Teeth: The Creation of United States Air Power for World War II* (Washington, D.C.: Smithsonian Institution Press, 1982), 15–16, 20–22; Herman O. Stekler, *The Structure and Performance of the Aerospace Industry* (Berkeley and Los Angeles: University of California Press, 1965), 8–12.
26. See the account in Ronald Fernandez, *Excess Profits: The Rise of United Technologies* (Reading, Mass.: Addison-Wesley, 1983), 155–57.
27. James W. Kunetka, *City of Fire: Los Alamos and the Birth of the Atomic Age, 1943–1945* (Englewood Cliffs, N.J.: Prentice-Hall, 1978); Fern Lyon, *Los Alamos, the First Forty Years* (Los Alamos, N.M.: Los Alamos Historical Society, 1984).
28. Markusen, Hall, Campbell, and Deitrick, *Rise of the Gunbelt.* See also the material on the formation of western cities and suburbs in Gerald D. Nash, *The American West Transformed: The Impact of the Second World War* (Bloomington: Indiana University Press, 1985); Robert Gottlieb and Peter Wiley, *Empires in the Sun: The Rise of the New American West* (Tucson: University of Arizona Press, 1985).
29. Markusen, Hall, Campbell, and Deitrick, *Rise of the Gunbelt,* chap. 8.
30. For detailed accounts of the Cold War–related growth of Los Angeles and Seattle, see Markusen, Hall, Campbell, and Deitrick, *Rise of the Gunbelt,* chaps. 5 and 7.
31. William G. Cunningham, *The Aircraft Industry: A Study in Industrial Location* (Los Angeles: Morrison, Oden, et al., 1951).
32. See, for instance, Rob Kling, Spencer Olin, and Mark Poster, eds., *Postsuburban California: The Transformation of Orange County since World War II* (Berkeley: University of California Press, 1991), 26ff.
33. Ernest Havemann and Patricia Salter West, *They Went to College* (New York: Harcourt, Brace, 1952), 234–35.
34. Mark Ellis, Richard Barff, and Ann Markusen, "Defense Spending and Interregional Labor Migration," *Economic Geography* 29:2 (1993), 1–22; Scott Campbell, "Interregional Migration of Defense Scientists and Engineers to the Gunbelt during the 1980s," *Economic Geography* 69:2 (1993), 204–223.
35. The following is based on Markusen and Yudken, *Dismantling the Cold War Economy,* 200–204.
36. Ann Markusen and Virginia Carlson, "Deindustrialization in the American Midwest: Causes and Responses," in Lloyd Rodwin and Hidehiko Sazanami, eds., *Deindustrialization in the U.S.: Lessons for Japan* (Boston: Unwin, Hyman, 1989), table 1.6.
37. See Ann Markusen, "Federal Budget Simplification: Preventive Programs vs. Palliatives for Local Governments with Booming, Stable, and Declining Economies," *National Tax Journal* 30:3 (1977), for a fuller exposition of this argument about the social costs of uneven development.
38. Not all regions that have shared in the military buildup vote consistently conservatively—New England is an example. But New England has not experienced a major

population increase; it has suffered net population outmigration over the entire post-war period and, even in the 1980s, the defense buildup was associated with lower population growth rates than the nation's and continued outmigration.

39. There are important exceptions, including Les Aspin and William Proxmire of Wisconsin, both of whom have, however, been more critical of Pentagon practices and requests than their gunbelt counterparts.

CHAPTER 3

The Illusion of Unity in Cold War Culture

Alan Brinkley

In 1987 the University of Minnesota hosted a lecture series on the Cold War in Amercian culture. Lary May, the organizer, wrote in his printed introduction to the subsequent book that "after 1945 Americans entered a new phase in their history" and that the United States had experienced "a paradigm shift of major proportions" in the framework of its culture.[1] He was not the first to make so bold a claim. In 1952 Mary McCarthy read a harsh critique of American society by Simone de Beauvoir and, characteristically, dashed off a sharp, somewhat overdrawn response:

> The society characterized by Mlle. De Beauvoir as "rigid," "frozen," "closed," is in the process of great change. The mansions are torn down and the real estate "development" takes their place: serried ranch houses, painted in pastel colors, each with its own picture window, and its garden, each equipped with deep freeze, oil furnace, and automatic washer, spring up in the wilderness. Class barriers disappear or become porous; the factory worker is an economic aristocrat in comparison with the middle-class clerk; even segregation is diminishing; consumption replaces acquisition as an incentive. The America invoked by Mlle. De Beauvoir as a country of vast inequalities and dramatic contrasts is ceasing to exist.[2]

Mary McCarthy was more defiantly jubilant (and more overtly polemical) than most commentators of her time. She was certainly more credulous of the claims of the new culture than later scholars have been. Yet she too saw the Cold War era as a fundamentally new time in American history. Her description captures some of the central qualities that many contemporary observers liked to attribute to American culture at the height of the Cold War, a society in which class divisions and fundamental social conflict were disappearing, in which even segregation was "diminishing." It was a culture that reflected an essential unity of interests and values widely shared by Americans of all classes, regions, races, and creeds.

If historical scholarship has done nothing else in the four decades since Mary McCarthy wrote those words, it has challenged and largely shattered the smug assumption that the United States in the 1950s—or at any time in its history—was a culturally unified nation. It has made visible the enormous range of social and cultural experiences within American society and the substantial, still unresolved conflicts that lay at its heart. But the fact remains that many Americans in the 1950s and the early 1960s believed in this effusive image of themselves. How was it possible for so many Americans to believe in something that now seems so clearly untrue? How did this illusion of unity become so important a part of American culture during the Cold War?

Critics of American culture in the 1950s and many scholars since have given much of the credit to the Cold War itself: to the political repression that accompanied the rivalry with the Soviet Union, to the pressures that rivalry created to celebrate American society and affirm its right to leadership of the "Free World." That was what Lary May meant in 1987 when he spoke of a "paradigm shift"—a new sense of the national self, driven in large part by the imperatives of the Cold War. Such observers are surely correct that the Cold War played a significant role in shaping the culture of its time. The official and unofficial repression of political belief, the pervasive fear among intellectuals and others of being accused of radical sympathies, the ideological fervor that the rivalry with the Soviet Union produced: all had a powerful effect on the way Americans thought about themselves and their culture and on what they dared do, say, and even think. It would be hard to overstate the degree to which the ideology and rhetoric of the Cold War shaped the public discourse of the time, hard to exaggerate the pervasiveness of its influence and the oppressiveness of its demands.[3]

Archibald MacLeish was one of a number of skeptical intellectuals of the time who noted with dismay the degree to which anticommunists had come to dominate and, he believed, corrupt American life. "Never in the history of the world," he wrote in 1949, "was one people as completely dominated, intellectually and morally, by another as the people of the United States were by the people of Russia in the four years from 1946 to 1949."[4]

And yet, hard as it is, many contemporaries and many scholars have overstated the role of the Cold War in shaping postwar American culture, indeed have suggested that it was virtually the only significant factor in shaping that culture. The Cold War provides a partial explanation of the character of postwar culture, but it is only one—and perhaps not the most important one—of several causes. Other social and cultural transformations had at least as much to do with the shaping of what we now call "Cold War culture" as did the Cold War itself.

The most obvious of these changes was the remarkable expansion of the American economy in the postwar years. It was, quite simply, the greatest and most dramatic capitalist expansion in American history, perhaps even in world history. It was often described at the time, not without reason, as the American "economic miracle." One economic historian, writing in the early 1960s, said of it: "The remarkable capacity of the United States economy represents the crossing of a great divide in the history of humanity." Through much of the twentieth century, particularly during the economic crises of the 1930s, substantial numbers of Americans had retained some skepticism about industrial capitalism. But during and after the war—as the prosperity rolled on and on and the new depression that so many had predicted in 1945 never came—it became possible for many Americans to believe that there were no limits, or at least no restrictive limits, to economic growth, that capitalism was capable of much greater feats than most Americans had once believed possible.[5]

The economist John Kenneth Galbraith, hardly an uncritical defender of capitalism through his long career, published a small book in 1952 entitled simply *American Capitalism.* In it he expressed some of the wonder and enthusiasm with which intellectuals and others faced this new discovery. About capitalism he had one succinct, almost breathless comment: "It works!" More than that, it obviated many of the principal dilemmas that had frustrated the reformers of earlier eras:

> In the United States alone there need not lurk behind modern programs of social betterment that fundamental dilemma that everywhere paralyzes the will of every responsible man, the dilemma between economic progress and immediate increase of the real income of the masses.[6]

Increasing the income of the masses, in other words, did not have to come at the expense of investment; the two things would reinforce one another. Poverty could be eliminated and social problems could be solved—not through the ideologically unattractive and politically difficult task of redistributing limited wealth, as many Americans had once believed would be necessary, but simply through growth. Or, as a member of the Council of Economic Advisers wrote in the late 1950s, somewhat more prosaically: "Far greater gains were to be made by fighting to enlarge the size of the economic pie than by pressing proposals to increase equity and efficiency in sharing the pie."[7]

A generation earlier economists, policymakers, and much of the public had despaired of ever seeing dynamic growth again. Many had talked of the arrival of a "mature economy" that had reached something close to the end of its capacity to grow, an economy whose fruits would have to be distributed more equally and perhaps more coercively given that those fruits seemed unlikely to expand. Now the same men and women were celebrating their discovery of the secret of virtually permanent economic growth. They were trumpeting the ability of economic expansion to solve social problems without requiring serious sacrifices, without the need for redistribution of wealth and power. When compared with what had come before, this was a paradigm shift of major proportions.

The belief that economic growth was the best route to a just society was part of a larger set of ideas, often called the "consensus" and associated most prominently with scholars and intellectuals. Consensus theorists promoted a set of ideas that, together, described America as a nation not only liberated from economic scarcity but also—and partly as a result—liberated from social conflict. That argument is associated most prominently with the sociologist Daniel Bell, the title of whose 1960 book *The End of Ideology* has become something of a label for American intellectual life in the 1950s. More revealing is the book's subtitle: *On the Exhaustion of Political Ideas in the Fifties.* In the absence of heated battles over scarce economic resources, Bell argued, Americans had run out of grand ideas:

> Ideology, which once was a road to action, has come to be a dead end. . . . Few serious minds believe any longer that one can set down "blueprints" and through "social engineering" bring about a new utopia of social harmony. . . . In the Western world, therefore, there is today a rough consensus among intellectuals on political issues: the acceptance of a Welfare State; the desirability of decentralized power; a system of mixed economy and of political pluralism. . . . The ideological age has ended.

And that, Bell concluded, was a good thing because, he wrote, "the tendency to convert concrete issues into ideological problems, to color them with moral fervor and high emotional charge, is to invite conflicts which can only damage a society." It was much better, in short, to live in a culture in which most people rejected great moral visions and broad social crusades and concentrated instead on the more prosaic, less exalted but also less dangerous business of competing for material advancement.[8]

The "end of ideology" idea helps explain why in the early and mid-1950s there was something of a decline in liberal activism, why many people had difficulty sensing great urgency in addressing social problems or launching new initiatives. For a time, at least, consensus ideology helped make American liberalism cautious, passive, even conservative. It was a worldview that sometimes seemed to rest on fear of what might happen if society embarked on any great crusades: a fear of class conflict, a fear of "mass irrationality, a fear of the dark passions that fanaticism could and had unleashed in the world." A distinguishing characteristic of the "consensus" liberalism was its fundamentally unradical, even antiradical quality.

Complacency would be too strong a word to describe the social outlook of consensus intellectuals. Many of them were harshly critical of American society and culture, even highly critical of the consensus itself. "A repudiation of ideology, to be meaningful," Daniel Bell wrote, "must mean not only a criticism of the utopian order but of existing society as well." The historian Richard Hofstadter, one of the first and most prominent spokesmen for the consensus idea, wrote scornfully in his classic 1948 study of a series of political leaders, *The American Political Tradition:*

> The sanctity of private property, the right of the individual to dispose of and invest it, the value of opportunity, and the natural evolution of self-interest and self-assertion . . . have been staple tenets of the central faith in American political ideologies. . . . American traditions also show a strong bias in favor of equalitarian democracy, but it has been a democracy in cupidity rather than a democracy of fraternity.[9]

Yet the concerns that Bell, Hofstadter, and other intellectuals expressed about the moral quality of the consensus did not alter their belief in the strength of the economic successes that supported it. Nor did it alter their essential (if slightly jaundiced) faith in the ability of the system to thrive. As they viewed their world, they concluded—to overstate things slightly—that there was no need to worry any longer about corporate power. Capitalism, after all, had proved that it worked. There was no need to worry about inequality; economic growth and social mobility would take care of that. Hofstadter, for example, wrote of the unemployed of the Great Depression: "The jobless, distracted, and bewildered men of 1933 have in the course of the years found substantial places in society for themselves, have become home owners, suburbanites, and solid citizens." Bell argued that "in a politico-technological world, property has increasingly lost its force as a determinant of power, and sometimes even of wealth. In almost all modern societies, technical skill becomes more important than inheritance as a determinant of occupation, and political power takes precedence over economic.

What then is the meaning of class?" The problems of American life, consensus theorists believed, were less those of inequality and injustice than of shallowness, banality, and alienation that modern culture produced.[10]

And yet, for all the criticism that consensus intellectuals often directed at their culture, there was also an unmistakably smug tone in some of the intellectual discourse of the 1950s: a tendency for writers to refer to their audience as "we," confident that the reading public was a homogeneous entity that shared their own values and assumptions. The great critic Lionel Trilling, for example, wrote with somewhat uncharacteristic ebullience that there comes a moment "when the tone, the manner and manners of one's own people become just what one needs, and the whole look and style of one's culture seems appropriate." Prosperity, abundance, consumerism, and the loosening of cultural prejudices that once would have barred Trilling, Bell, and other Jews from a place at the center of American life shaped this new view of the nation's culture. The Cold War simply reinforced it.[11]

These ideas—the assumption of increasing and virtually universal abundance, the assumption of shared values and goals, the belief in the end of conflict—reflected the experiences of members of the white middle class and of educated white middle-class men in particular. That suggests another set of changes of considerable importance to the shaping of postwar culture: changes in the size and character of the American bourgeoisie.

Definitions of the middle class are subject to dispute. Yet by almost any reasonable definition, the American middle class was expanding dramatically in the postwar years. It was expanding occupationally. In 1956 for the first time in American history, government statistics showed that white-collar workers outnumbered blue-collar workers in the United States. It was expanding economically through a rapid growth in the number of people able to afford what the government defined as a "middle-class" standard of living. In 1929, 31 percent had achieved that standard. In 1955, 60 percent had achieved it. It was expanding educationally. The percentage of young people graduating from high school rose from just under 47 percent in 1946 to over 63 percent in 1960, and the percentage of young people attending college rose from 12.5 to 22 over the same period. In terms of consumption patterns, things traditionally considered middle-class attributes were becoming more common. Home ownership rose from 40 percent in 1945 to 60 percent in 1960. By 1960, 75 percent of all families owned cars; 87 percent owned televisions; 75 percent owned washing machines. Owning such things did not by itself make someone middle class, any more than attending college or holding a white-collar job did. But such changes in the material conditions of life tended to transform the self-image of many people, helping them to consider themselves part of the great middle class, beneficiaries of what has now come to be called the "American Dream" of education, home ownership, material comfort, and economic security. They were the products of what the sociologist William Whyte called the "second great melting pot," a process of socialization that helped men and women transcend not ethnicity but class, that created a new and more pervasive American bourgeoisie. The newcomers to the middle class, Whyte wrote, "must discard old values" and adopt new ones.[12]

The middle class was not simply becoming more numerous in these years. It was also becoming more homogeneous and more self-conscious. One reason for this was the growing pervasiveness of middle-class images, ideas, and values in American popular culture, a result in part of changes in the media in the 1950s. The most important of those changes was the emergence of television, which within a decade moved from being a curiosity to being a central fact of life for virtually everyone in America—the most powerful force in American culture. Far more than newspapers or magazines or even radio or movies, television linked society together and provided a common cultural experience.

What, then, was that cultural experience? What message did television convey? There were many messages, then as now. But after a brief period in the early 1950s of diverse and innovative programming, television began to succumb to its own economic imperatives and for many years studiously avoided controversy and conflict. It offered instead a relatively homogeneous image of American life, dominated by middle-class lifestyles and middle-class values. This was, to use the Marxist phrase, "no accident."

Television programming in the 1950s was dominated by a very few people—the executives of the three major networks and the commercial sponsors they were attempting to attract. It was a very different programming world from the one Americans came to know in later decades. The power of sponsors over television programming today is relatively limited. Most shows are produced long before sponsors are even approached; advertisers simply buy time slots and only rarely have anything to say about program content. But in the 1950s the network had to court advertisers. Each program was generally supported by a single sponsor who exercised considerable and often direct and prior control over program content.

Many televsion programs in the 1950s actually bore the names of their sponsors: the Pepsi Cola Playhouse, the General Electric Theater (whose host, Ronald Reagan, helped launch his political career through his identification with the company and its programming), the Dinah Shore Chevy Hour, Alcoa Presents, the Camel News Caravan, and others. Corporations whose identities were tied up so directly with particular programs were reluctant to permit them to become controversial, divisive, or even unusual. That was a reflection in part of the advertising assumptions of the 1950s: effective advertising tried to appeal to everyone and to alienate no one. Network executives believed, probably correctly, that to attract sponsors they needed to provide programs that were consistently uncontroversial. One executive, responding to a proposal for a television series dealing with problems of urban life, wrote: "We know of no advertiser or advertising agency in this country who would knowingly allow the products which he is trying to advertise to the public to become associated with the squalor and general down character of this program."[13]

From such assumptions emerged the characteristic programming of the late 1950s and early 1960s: westerns, variety shows, quiz shows, and above all situation comedies, the quintessential expression of the middle-class view of American life. Popular situation comedies—The Adventures of Ozzie and Harriet, Father Knows Best, Leave It to Beaver, Dennis the Menace—were set in virtually interchangeable suburban houses, with virtually interchangeable families, in virtually interchangeable situations. Almost everyone was upper middle class. Almost everyone was white. Almost everyone lived in a stable nuclear family. Most situation comedies reinforced prevailing notions of gender roles: women did not work outside the home, or if they did they were unmarried and were working principally as secretaries and teachers. Men left the house in the morning to go to nameless white-collar jobs in the city.

The world of television entertainment programming was, with only a very few exceptions, a placid, middle-class world. Even the exceptions—shows such as The Honeymooners or I Love Lucy or The Life of Reilly—which revealed elements of working-class or ethnic life or presented women as something more than contented housewives, took images of difference and domesticated them, so that in the end they too reinforced rather than challenged assumptions about the universality of middle-class ideas and experiences. Middle-class Americans, seeing such constant confirmation of their own world on television, could easily conclude that this was the world in which virtually all Americans lived.[14]

Another development that played an important role in shaping the distinctive outlook of the American middle class was the rise of the new, or at least newly expansive, suburban culture in the 1950s. Suburbanization isolated many Americans from the diversity and abrasiveness of urban life. It also provided them with what were at first stable and relatively homogeneous communities. All suburbs were not alike; but within most suburbs, particularly within many of those that grew up in the 1950s, there was a striking level of uniformity and conformity. The most obvious effect of suburbanization, therefore, was a standardizing of the outward lives of those who lived in the suburbs.[15]

This was visible even in the architecture of suburbia. Most suburban developments tended to be built all at once, by a single developer, often designed by a single architect or no architect at all, many characterized by similar and often identical houses. The most famous examples of this are the Levittowns, which became models for other, relatively inexpensive suburban developments. Relatively few suburbs were built by the same kind of mass production that the Levitt family pioneered for their developments. But many were built in ways that produced a similar homogeneity. Even the more expensive suburbs used elaborate zoning and building codes to ensure that homes would not diverge too radically from the community norm. In almost all suburbs, homes were designed to thrust the focus of the family inward on itself, not outward into the community. Suburbanites used their back yards, not their front yards, for recreation. They built back, not front, porches. They valued privacy more than interaction with the neighborhood.[16]

There was a cultural uniformity as well within many suburbs. Sociologists and others who studied suburban communities in the 1950s were often highly critical of the overpowering conformity they found there. David Riesman wrote, "The suburb is like a fraternity house at a small college in which likemindedness reverberates upon itself." William Whyte, who studied a suburb outside Chicago, found there what he called "a belief in 'belongingness' as the ultimate need of the individual." Riesman and Whyte were only half right. Suburbs did not create a pressure to conform as much as they were products of a desire to conform—a desire of men and women to gather in communities of likeminded people, of common class and often common ethnic and religious characteristics. But whatever the reasons, suburbs insulated their residents from social and cultural diversity.[17]

The suburban population as a whole was highly diverse—economically, ethnically, even racially. There were working-class suburbs, ethnic suburbs, black suburbs. Yet few individual suburban communities were diverse; few were places where white Anglo Americans lived alongside African Americans or Hispanic Americans or other minorities. One of the reasons for the massive movement of middle-class whites into suburbs—although not the only reason—was the desire to escape the racial and class heterogeneity of the cities. "Suburbia . . . is classless," William Whyte wrote in 1956, "or at last its people want it to be." There was in most suburbs, another social critic observed, "no elite, no wealthy, prestigious upper class. There were no shanty families, no clusters of the ethnically 'undesirable.' "

But in the larger world, of course, there were wealthy elites, there were poor people, and there were racial and ethnic minorities. The growth of suburbs did not reduce their numbers. Instead it protected the middle class from contact with them, even from active awareness of them. It was not so much a force for homogenizing American society at large as it was a force for dividing it. But in the process it helped make possible an increasingly uniform middle-class culture and an increasingly common middle-class view of the world.[18]

Another of the large social forces that were forging a distinctive middle-class culture in the postwar era was the increasing bureaucratization of white-collar work, the growing proportion of middle-class men and a slowly increasing number of middle-class women whose lives were embedded within large-scale corporate and government organizations that created pressures of their own for conformity and homogeneity. Some of the same social critics who attacked the suburbs as stifling and alienating launched similar attacks on the "organization." Employees of large corporate organizations, according to their critics, were becoming something close to automatons. They were pressured to dress alike, to adopt similar values and goals and habits, to place a high value on "getting along" within the hierarchical structure of the corporation. The organization, its critics argued, posed a challenge to the capacity of individuals to retain their psychological autonomy. It was creating alienated conformists afraid to challenge prevailing norms; people who would take no risks; people who feared to be different.[19]

In his 1956 book *The Organization Man,* William Whyte criticized the bureaucracy in much the same way he criticized the suburb. Corporate workers, he argued, faced constant pressures to get along by going along; they were victims of a social ethic "which makes morally legitimate the pressures of society against the individual." David Riesman's 1950 book *The Lonely Crowd,* one of the most influential works of postwar sociology, argued that the modern organizational culture was giving birth to a new kind of individual. In earlier eras, most men and women had been "inner-directed" people, defining themselves largely in terms of their own values and goals, their own sense of their worth. Now the dominant personality was coming to be the "other-directed man," defining himself in terms of the opinions and goals of others, or in terms of the bureaucratically established goals of the organization. This new form of character was blind to distinctions of class, Riesman argued. "Both rich and poor avoid any goals, personal or social, that seem out of step with peer-group aspirations." And it was debilitating to true freedom. "Men are created different," he lamented; "they lose their social freedom and their individual autonomy in seeking to become like each other."[20]

The sociologist C. Wright Mills offered a more radical critique, which described the way in which modern organizations created dangerous centers of power, remote from popular will. Later, that helped make his books—*The*

Power Elite and *White Collar*—influential documents within the New Left. But Mills's work was at least as notable for its devastating portrait of the way in which large organizations stifle the individuality of those who work in them, both by creating expectations of conformity and by exercising coercive authority through remote and inaccessible systems. This kind of power was more dangerous and more difficult to challenge than even the most arbitrary forms of authority in older, less bureaucratized social settings. "In a world dominated by a vast system of abstractions," he wrote, "managers may become cold with principle and do what local and immediate masters of men could never do. Their social insulation results in deadened feelings in the face of the imprisonment of life in the lower orders and its stultification in the upper circles."[21]

The stifling uniformity of modern suburban and organizational life was a common theme in the work of literary figures of the 1950s. Writers such as John Cheever, Norman Mailer, John Updike, Saul Bellow, J. D. Salinger, and Walker Percy wrote novels that centered around lonely, frustrated, white, middle-class, male protagonists struggling to find some way to bring meaning and fulfillment to empty, rootless, unsatisfying lives. The Beat poets whose critique of modern society was far more sweeping and more radical in its implications than those of mainstream middle-class writers nevertheless shared their fear of the stifling quality of bureaucratic life. "Robot apartments. Invincible suburbs. Skeleton treasuries. . . . Spectral nations," Allen Ginsberg wrote in his searing poem *Howl*, which became an anthem of the Beat generation. These were not new themes in the 1950s, certainly. But they were newly directed toward suburban life, toward the corporate work place, toward the facelessness and homogeneity of bourgeois society.[22]

Such critiques are striking in retrospect because they almost entirely overlooked what later history has made clear was the most frustrated group within the middle class: not the men but the women. They are striking, too, because they say very little about the vast numbers of Americans outside the middle class altogether, barred from its successes either by economic circumstance or by active barriers of discrimination. And they are striking because almost nowhere in these diagnoses of the character of middle-class society, or of the angst it created among some of its members, was there any significant discussion of the Cold War. What shaped the world of the American bourgeoisie, both its critics and its celebrants were suggesting, were the cultural, economic, and demographic forces of a rapidly evolving industrial society and only incidentally the pressure of the struggle against communism.

The smooth surface of postwar middle-class culture—and the discontents festering below its surface, which would in the 1960s challenge and even shatter it—parallel the smooth surface of postwar American foreign policy and the critiques that would shake it, too, in the 1960s. American culture and American foreign policy reinforced one another in countless ways in the age of the Cold War. Yet they did not cause one another. American society and culture would likely have looked much the same in the 1940s and 1950s with or without a Cold War.

Yet the Cold War remains a powerful metaphor for describing that culture. The architects of the Cold War came to view a diverse and rapidly changing world through the prism of a simple ideological lens, smoothing out the rough spots and seeing a uniformity of beliefs and goals that did not in fact exist. The architects of postwar middle-class culture looked at a diverse and rapidly changing society in the United States through a similarly limited, self-referential perspective. They constructed and came to believe in an image of a world that did not exist. America did, as Mary McCarthy and many others claimed at the time, enter a "new era in its history" in 1945. But it was not the era they thought it was, and it did not produce the history they expected.

Notes

1. Lary May, "Introduction," in Lary May, ed., *Recasting America: Culture and Politics in the Age of the Cold War* (Chicago: University of Chicago Press, 1989), 14.
2. Jackson Lears, "A Matter of Taste: Corporate Cultural Hegemony in a Mass-Consumption Society," in May, *Recasting America,* 38–39; Mary McCarthy, "Mlle. Gulliver en Amérique," *The Reporter* (January 22, 1952), 36.
3. See, for example, Stephen J. Whitfield, *The Culture of the Cold War* (Baltimore: Johns Hopkins University Press, 1991).
4. Archibald MacLeish, "The Conquest of America," *Atlantic Monthly* (1949; reprinted March 1980), 37–38.
5. Godfrey Hodgson, *America in Our Time* (Garden City, N.Y.: Doubleday, 1976), 78–83; William Chafe, *The Unfinished Journey: America since World War II* (New York: Oxford University Press, 1995), 112.
6. Kenneth Galbraith, *American Capitalism: The Concept of Countervailing Power* (Boston: Houghton Mifflin, 1952), 12–16.
7. Hodgson, *America in Our Time,* 81.
8. Daniel Bell, *The End of Ideology: On the Exhaustion of Political Ideas in the Fifties* (New York: Free Press, 1960), 393–94, 402–3.
9. Bell, *End of Ideology,* 16; Richard Hofstadter, *The American Political Tradition and the Men Who Made It* (New York: Alfred A. Knopf, 1948), viii.
10. Richard Hofstadter, "The Pseudo-Conservative Revolt (1955)," in Daniel Bell, ed., *The Radical Right* (Garden City, N.Y.: Doubleday, 1963), 75; Bell, *End of Ideology,* 398. Among other important statements of the consensus idea are Louis Hartz, *The Liberal Tradition in America* (New York: Harcourt, Brace & World, 1955), and David Potter,

People of Plenty: Economic Abundance and the American Character (Chicago: University of Chicago Press, 1954).

11. Lionel Trilling, *A Gathering of Fugitives* (Boston: Beacon Press, 1956), 67–76; Thomas Bender, "Lionel Trilling and American Culture," *American Quarterly* 42 (1990), 340–41.
12. Bureau of the Census, *Historical Statistics of the United States* (Washington, D.C.: Government Printing Office, 1975), I:289–310, 379, 383; William H. Whyte Jr., *The Organization Man* (New York: Simon & Schuster, 1956), 299–300; James T. Patterson, *Grand Expectations: The United States, 1945–1974* (New York: Oxford University Press, 1996), 311–42; Richard Polenberg, *One Nation Divisible: Class, Race, and Ethnicity in the United States since 1938* (New York: Viking Press, 1980), 129–39; Chafe, *Unfinished Journey,* 111–22.
13. James Baughman, *The Republic of Mass Culture: Journalism, Filmmaking, and Broadcasting in America since 1941* (Baltimore: Johns Hopkins University Press, 1992), 48–75; Hodgson, *America in Our Time,* 137–52.
14. Karal Ann Marling, *As Seen on TV: The Visual Culture of Everyday Life in the 1950s* (Cambridge, Mass.: Harvard University Press, 1994), 129–35, 210–17; Patterson, *Grand Expectations,* 348–55.
15. Kenneth T. Jackson, *The Crabgrass Frontier: The Suburbanization of the United States* (New York: Oxford University Press, 1985), 231–35.
16. Herbert Gans, *The Levittowners: Ways of Life and Politics in a New Suburban Community* (New York: Alfred A. Knopf, 1967), 3–44; Zane L. Miller, *Suburb: Neighborhood and Community in Forest Park, Ohio, 1935–1976* (Knoxville: University of Tennessee Press, 1981), 46–96.
17. David Riesman, "The Suburban Sadness," in William M. Dobriner, ed., *The Suburban Community* (New York: G. P. Putnam), 375–402; Whyte, *Organization Man,* 32–46, 330–81; Polenberg, *One Nation Divisible,* 133–37.
18. Whyte, *Organization Man,* 298, 307–10; Polenberg, *One Nation Divisible,* 141.
19. See Daniel Horowitz, *American Social Classes in the 1950s* (Boston: Bedford Books, 1995), 1–26; Whyte, *Organization Man,* 63–167.
20. Whyte, *Organization Man,* 392–400; David Riesman, *The Lonely Crowd: A Study of the Changing American Character* (New Haven: Yale University Press, 1950), 13–23, 306–7.
21. C. Wright Mills, The Power Elite (New York: Oxford University Press, 1956); C. Wright Mills, *White Collar* (New York: Oxford University Press, 1951), 110–11.
22. Gene Feldman and Max Gartenberg, eds., *The Beat Generation and Angry Young Men* (New York: Citadel Press, 1958), 164–74.

CHAPTER 4

"We'll Follow the Old Man"

The Strains of Sentimental Militarism in Popular Films of the Fifties

Christian G. Appy

> We'll follow the Old Man wherever he wants to go,
> Long as he wants to go opposite to the foe.
> We'll stay with the old man wherever he wants to stay,
> Long as he stays away from the battle's fray.
> Because we love him, we love him . . .
> The grandest son of a soldier of them all.[1]

Christmas Eve, 1944. In the European "theater," amid ruins of war, American troops enjoy a holiday revue. Artillery shells whistle and thump in the distance. A new commander arrives. He is appalled by the breach of discipline: "These men are moving up tonight, General Waverly. They should be lined up for full inspection." The Old Man feigns agreement: "You're absolutely right. . . . There's no Christmas in the army." But this is just tough talk. The kind-hearted general sends his kill-joy replacement on a detour, allowing his men a chance to hear Bing Crosby sing "White Christmas," the sentimental, homesick fantasy of bygone days that became the most popular song of the 1940s. The show ends with another Irving Berlin tune, a "slam-bang finish" in honor of General Waverly, "The Old Man."[2]

So the 1954 film musical *White Christmas* begins and ends. "The Old Man" and "White Christmas" are reprised at film's close, neatly tucking the postwar story inside a holiday card from the sentimental heart of history's bloodiest war. The use of World War II as frame, setting, or flashback characterizes many Cold War fictions of the 1950s and serves a variety of functions. Often, as in *White Christmas,* the "Good War" provides a blanket of moral certitude and nostalgia under which Cold War militarization is hidden, justified, or comforted. The earlier war also offers a model of narrative drive and closure for a period notable for its plotlessness and lack of finality. World War II can open small windows of resistance to Cold War orthodoxy. Even some flag-waving films of the 1950s such as *The Bridges at Toko-ri* and *Strategic Air Command* are haunted by a sense of decline from the unity and purpose of the prior war. Reluctance and doubt leak through even as they are denounced. And stories such as *The Caine Mutiny* and *Mister Roberts* are "about" World War II but deal with issues of loyalty and undramatic military duty that are more pertinent to the Cold War.[3]

The martial cadence of "The Old Man" might lead a casual listener to suspect it had been lifted directly from Irving Berlin's popular World War II musical *This Is the Army.* In 1942 Berlin called Gen. George Marshall for official permission to make an all-soldier, morale-boosting revue. With the military's sanction, *This Is the Army* soon became a Broadway hit featuring a cast of 359 soldiers. Quick to recognize the show's global potential, Berlin took his troupe of troops on a national and international tour that lasted until 1945 and reached two-and-a-half million soldiers and civilians. A film version of *This Is the Army* (1943) became Hollywood's greatest box-office success from 1943 to 1949.[4]

Its popularity was short-lived. Today it strikes most viewers as a historical curiosity and its video sales are negligible. Other wartime film musicals such as *Holiday Inn* (1942) have, over time, drawn many more viewers than *This Is the Army.* Why was the soldier review so popular? In part its appeal was the sheer quantity and variety of pleasures it offered—weepy love ballads, patriotic anthems, campy drag numbers, black face "Mandy," vaudevillian bits (jokes, jugglers, impressionists, acrobats), an all-black dance number, and innocuous ditties about the rigors of army life. Prior to television's ascendancy, even a virtually plotless cinematic variety show could attract a mass audience. James Agee picked up on a central element of how the film connected with audiences: "A simple-hearted friendliness generated between audience and screen at *This Is the Army* made that film happy to see even when it was otherwise boring."[5] With dozens of audience reaction shots as the "stage" play unfolds, film viewers were invited to share the growing bond between civilians (the "audience") and soldiers (the "actors"). That many of the actors were, in fact, amateur performers—joined by Hollywood contract players like Ronald Reagan and George Murphy—further erased the distinction between spectator and participant. The film imaginatively united civilian and soldier, homefront and warfront.

This Is the Army was intended as unabashed pro-war propaganda and some of its numbers are explicitly militaristic (e.g., "More bombers to attack with/More bombers till the skies are black with—American Eagles!"). Since at least the 1960s such songs have drawn laughs, or disgust, from many audiences and they were cleverly excerpted by Peter Davis in his anti–Vietnam War documentary *Hearts and Minds* (1974). But *This Is the Army* really only has a few blatantly bellicose scenes and those are surrounded by a thick layer of sentimentality. The soldiers are represented more as mommas' boys than ruthless killers.[6] The film's major propositions are as banal as this: (1) U.S. soldiers are peace-loving citizen soldiers, (2) the fight is on behalf of women and loved ones at home, (3) combat is an unwelcome but quintessential expression of manhood and patriotism, and (4) the military is a melting pot of democratic values. *This Is the Army* is a classic work of sentimental militarism.

With roots that could go back centuries, sentimental militarism may have achieved its fullest form—or its last unequivocal gasp—in American popular culture of the Second World War. That said, cultural historians should examine the ways sentimental militarism has persisted and transformed itself throughout the Cold War. While both World War II and its aftermath produced many grim representations of American life (e.g., film noir), it would be hard to exaggerate the power and persistence of sentimentality in American culture. What most interests me here is the strained and defensive tone that characterizes much of the sentimental militarism produced during the early Cold War, even well before the 1960s when the peace movement challenged all dominant expressions of American militarism.

Even *White Christmas* marks some significant breaks from *This Is the Army.* While it shares much of the earlier musical's mix of humor, romance, and flag-waving, the central "martial" tune in *White Christmas*—"The Old Man"—is curiously conditional. In *This Is the Army,* the soldiers sing "we're dressed up to win" without irony, however funny it may be today. "The Old Man" invites a more ambiguous response. It begins with a claim of unqualified loyalty—"we'll follow the old man, wherever he wants to go." But the claim is immediately subverted—"as long as he wants to go, opposite to the foe." Loyalty, it turns out, is predicated on the general's willingness to keep his men out of combat. No one suggests that the general himself is shy of battle. As Bing Crosby says when introducing the song: "The old man's moving toward the rear. That's a direction he's never taken in his life." Moreover, the men claim to love their commander for instilling discipline (for keeping them "on the ball"), an appreciation already turning nostalgic ("We'll tell the kiddies we answered duty's call . . ."). "The Old Man" has it both ways: it extols military authority and discipline even as it seeks to avoid the "battle's fray." As much as the men love General Waverly, they no longer want to follow him into righteous battle (like the troops in "This Is the Army") but to follow their leader into military retirement.

White Christmas was not as popular as *This Is the Army,* but it was a sure-fire holiday hit from Paramount. Good box-office was virtually assured by a four-million-dollar budget and a team of veteran performers—Bing Crosby, Danny

Kaye, Rosemary Clooney, Vera-Ellen, Dean Jagger, director Michael Curtiz *(This Is the Army, Mission to Moscow, Casablanca, Mildred Pierce),* and songwriter Irving Berlin, age 65. And if that were not enough, Paramount announced that this would be the first film shot with their new wide-screen technique called VistaVision.[7]

A hit it was, but critical praise was grudging at best. *White Christmas* had a few decent new Berlin tunes, most critics conceded, and a fine cast but most of the material, like Berlin, was getting stale. The plot was "dated" (*Newsweek*), "simple" (*Life*), "sentimental" (*Time*), "foolish" (*Commonweal*), and "like a re-make of virtually every backstage musical produced during the Thirties" (*Saturday Review*). Some critics attributed the film's old-fashioned feel to the fact that its setting, title song, and a few plot ideas were drawn from *Holiday Inn* (1942).[8]

Derivative and nostalgic as it is, *White Christmas* has maintained a popularity during the last 50 years that has far surpassed *This Is the Army* and even the critically more acclaimed *Holiday Inn. White Christmas* has survived, I believe, partially because it provides one of the most appealing Cold War versions of sentimental militarism, a constellation of attitudes and values that were central to the way dominant American culture denied, evaded, and justified the actualities of its increasingly militarized society.

It represents through erasure and historical sentimentality a cultural accommodation to the development of a permanent national security state. By the time *White Christmas* appeared in 1954, America had initiated a global policy of anticommunist containment and counterrevolution, had developed a national security system that included massive public and private efforts to eliminate left-wing domestic dissent, had instituted the first peacetime draft, had used the CIA to build, support, and sometimes overthrow governments around the world, had committed itself to a permanent wartime economy, had fully embarked on a nuclear arms race and an enormous program of domestic civil defense, and had just concluded a brutal and stalemated war in Korea.[9] Not all of these developments were uncontested, especially in the 1940s, but most Americans were either largely unaware of the transformations (the CIA's operations in particular), or had a set of ideas about American society that translated the new realities into legitimizing stories. Hollywood played a crucial role in providing these translations.

Like *White Christmas,* most popular Hollywood films made little if any direct reference to Cold War realities. The most explicitly political and anticommunist films tended to be box office flops.[10] No wonder: most were as tedious as they were unconvincing. The buffoonish anticommunist father in *My Son John* (Dean Jagger) must have struck many viewers—especially young people—as more monstrous than his sly but sometimes funny communist son (Robert

Walker).[11] So it is probably wise to turn to popular film as a better vehicle for exploring the cultural work of encouraging Americans to follow the "old men" into an age of Cold War militarization while believing that their nation had remained essentially peace-loving and antiauthoritarian. Though it is true that film audiences were declining from a postwar high of 90 million viewers per week (1946–1948) to 40–50 million in the 1950s, the most popular films of the mid-50s were still "the mass medium of choice for a heterogeneous, multigenerational audience."[12]

No film is an unambiguous index of popular values. Nor are popular films simplistic mouthpieces of hegemonic forces. To appeal broadly, they must open themselves to alternative interpretations and expose inconsistencies and contradictions.[13] But this very complexity makes them all the more valuable in interpreting the dynamics of ideological persuasion and reception. As much as any cultural construction, Hollywood films expose one of the most fascinating paradoxes of the early Cold War. On one hand, Cold War assumptions about American virtue and communist evil were a surprisingly easy sell—widely shared and rarely challenged. On the other hand, efforts to dramatize those assumptions often seem overdetermined, plotless, and surprisingly unconvincing. This paradox may be characteristic to one degree or another of all orthodoxy, yet Cold War narratives lacked the narrative coherence and appeal of earlier American war stories.[14]

This chapter focuses on six popular films that opened in the mid-1950s: *White Christmas* (1954), *The Caine Mutiny* (1954), *Mister Roberts* (1955), *The Bridges at Toko-ri* (1954), *Strategic Air Command* (1955), and *The Man in the Gray Flannel Suit* (1956). Each of these stories is multitextual. *White Christmas* was built on Irving Berlin songs that had an impact transcending the screen, and all of the other films, with the exception of *Strategic Air Command,* were based on best-selling novels. In addition, two of the stories, *The Caine Mutiny* and *Mister Roberts,* had successful runs as Broadway plays. These were, in fact, among the most popular fictions of the early Cold War. The popular texts I've chosen are constructed in various ways as defenses of dominant American institutions and policies. And I believe that they were largely successful in reinforcing the kind of assumptions and attitudes that gave public sanction to official Cold War policy. Yet these popular fictions appealed as well to more troubled responses to the era and reveal cracks in the dominant ideology not easily erased or evaded. While World War II succeeded in producing some classic works of sentimental militarism, these formulas could not be transferred to the Cold War without revealing significant strains.

White Christmas: Deference and Denial

White Christmas moves quickly from war to a prosperous peace.[15] After General Waverly (Dean Jagger) is sung into retirement, we go to a Florida nightclub where wartime crooners Bob Wallace (Bing Crosby) and Phil Davis (Danny Kaye) are performing. The successful song-and-dance team of Wallace-and-Davis has made a fast cut to fame and fortune but we only get to bask in the glamour for one medley before some minor discontent arises. Phil wants more time away from Bob, his workaholic partner, and urges him to marry. Bob's response: "The kind of girls you and I meet in this business—they're young, they're ambitious, they're full of their own careers. They're not interested in getting married, settling down, raising a family." And they're stupid—"they can't even spell Smith" [College].

Seconds later the right kind of girls appear in the form of a "sister act"—the "Haynes Sisters" (Rosemary Clooney and Vera-Ellen). Phil manages to get all four entertainers on a train to Vermont where the Haynes Sisters are booked at a ski lodge. But Vermont has no snow, and the inn has no guests. The bigger surprise is that the lodge is owned by the men's former commander, General Waverly. He "sank everything" into the failing venture. Since the Old Man is as uncomplaining as ever, songwriter Irving Berlin must dramatize the awful plight of retired generals:

> When the war was over there were jobs galore
> For the GI Josephs who were in the war.
> But for Gen'rals things were not so grand. . . .
> Who's got a job for a General, when he stops being a General?
> They all get a job, but a General no one hires.[16]

Did postwar audiences give this sentiment any literal credence? Probably not. Surely experience itself had demonstrated the advantages officers had over enlisted men in establishing postwar careers. Yet many moviegoers were emotionally disposed to believe there were "old soldiers" who deserved greater appreciation. After all, many had wept just a few years before when General MacArthur lost his job. It seemed to require hyperalert critics like C. Wright Mills to point out that old soldiers like MacArthur did not "fade away" or even "retire" but exchanged their high military rank for positions of wealth and power in the corporate sector of the military-industrial complex. General Waverly may have needed a fund-raiser, but his colleagues were cashing in.[17]

To bolster the general's business, Wallace-and-Davis bring their whole production company to Vermont and make a television plea for all veterans from the Old Man's division to show up for a holiday reunion at the inn. The entertainers cover the whole cost: "We're not commercializing on the Old Man's hard luck!"

As the inn fills with veterans, a housekeeper's ruse requires General Waverly to wear his old full-dress military uniform.[18] The general enters the hall to find his former soldiers at attention and ready for inspection. He obliges the ritual with mock severity:

> I am not satisfied with the conduct of this division. Some of you men are under the impression that having been at Anzio entitles you not to wear neckties. Well you're wrong. . . . And look at the rest of your appearance. You're a disgrace to the outfit. You're soft, you're sloppy, you're unruly, you're undisciplined. And I never saw anything look so wonderful in my whole life.

This moment has the capacity to wring tears from a surprising variety of viewers—young and old, men and women, liberal and conservative. It is the high-water mark of postwar sentimental militarism. Military uniforms, military service, military authority are all passionately esteemed. Yet equally celebrated as "wonderful" is the sloppy peacetime warmth of the veterans. Militarism is located in the past, not as an ongoing and intensifying characteristic of postwar reality. Sentimental militarism denies its reverence of permanent militarization even as it celebrates the love of military authority that is at least one precondition for widespread acquiescence to future wars.

The finale of *White Christmas* invites pleasure in the status of veteran: a whole nation of veterans and their families pay homage to the military while enjoying peacetime. The film makes no demands on its audience to continue "serving" or "sacrificing." Shot just after the Korean War, it makes no reference to any military event after World War II. There is only one indication that the military even continues to exist. The moment is telling. General Waverly, his civilian business floundering, applies for reactivation as commander of a combat unit. The military turns him down. We are invited to feel sympathy for this blow to "old Tom's" ego, but he betrays no bitterness, concluding that it might be time to take up horseshoes after all. The film also includes a campy number called "Gee, I Wish I Was Back in the Army" (performed as country bumpkins), making clear that no one but the general harbors serious fantasies of re-enlistment. Whether re-activated or retired, we are confident that Waverly is precisely the kind of Old Man who could lead the nation in the postwar world. Other, younger veterans, can laugh at their beer bellies, but the Old Man is as trim as ever, strapping in his dress uniform.

The life and music of Irving Berlin is central to the extra-textual significance of *White Christmas.* Born in 1888, as Israel Baline, Berlin was the sixth child of a cantor who died early in Berlin's life. From an impoverished childhood on the lower east side of New York, Berlin was already famous by World War I. He was a gifted assimilator of a variety of musical traditions and an equally eager champion of dominant American institutions and values. *White Christmas* is a supreme example of Berlin's attraction to a vision of America devoid of all evidence of fundamental difference and conflict. Under this sentimental snow, Berlin suppresses his own Jewish heritage and provides America with its most effective assimilationist Christmas carol: a song that gives cultural primacy to the Christian holiday by stripping it of any obvious theological or exclusionary implications.

Berlin regarded himself, and was widely regarded, as a great patriot. He donated all his royalties from "God Bless America" to the Boy Scouts and Girl Scouts of America and all the earnings from "This Is the Army" to the Army Emergency Relief Fund. He also had a lifelong reverence for military leaders, and gave his political support to Dwight Eisenhower. Berlin wrote "They Like Ike" for the Broadway show *Call Me Madam* which opened in October 1950. He wrote a revised version called "I Like Ike" for an Eisenhower fund-raiser in February 1952. The networks banned Berlin's song after Eisenhower announced his candidacy, but Eisenhower's media advisers used the song, or jingles based on it, at campaign stops, the Republican convention, and on televised commercial spots. Two years later Berlin wrote "I Still Like Ike," and in 1956 he added "Ike for Four More Years."[19] In many respects, *White Christmas* may be viewed as a re-election campaign film for Eisenhower. In 1954, many viewers must surely have identified the grandfatherly General Waverly with President Eisenhower. Dean Jagger captures a great deal of that strong but approachable friendliness that characterized Eisenhower's public persona as president. And Eisenhower's enormous love of recreation (including hunting, fishing, bridge, painting, and—during his presidency alone—some eight hundred rounds of golf) certainly evoked an image of semiretirement, more like Dean Jagger running a country inn than a former commander seizing the reins of power.[20]

The sentimental militarism of *White Christmas* is located in its simultaneous celebration of military authority and its utter denial that this authority exercised an increasingly potent force in American life. It reflects Tom Engelhardt's acute observation that one peculiar aspect of postwar militarization in U.S. political culture was that the public seemed to be simultaneously mobilizing and demobilizing, preparing on one hand for a dreadful apocalypse and on the other for an arcadia of abundance.[21] Moreover, since many of the actions of the national security state operated in secret, Americans could remain essentially oblivious to its arrival. Fictions like *White Christmas* helped to secure the veil. It courted tacit support for the rise of a distant, largely invisible militarized state by inviting Americans "away from the battle fray" to the hills of Vermont, to a world of leisure, holiday romance, and matrimony where the racially homogeneous population guarantees a white Christmas whether it snows or not.[22]

This form of sentimental militarism, however, rested on an erasure of Cold War reality that was extreme even for Hollywood. In an age of loyalty oaths, civil defense drills, the Berlin airlift, atomic spy cases, the Chinese Revolution, the Korean War, and the Army-McCarthy hearings, many films suggested that Dean Jagger's general was, like his name, much too Waverly.[23] Waverly himself endorses his spit-and-polish replacement: "He's tough. Just what this sloppy outfit needs. He'll have you standing inspection night and day. You may even learn how to march. And if you don't give him everything you've got I may come back and fight for the enemy."

What might happen if Tom Waverly were replaced by Philip Queeg?

The Caine Mutiny: A Strained Tribute to "Constructive Loyalty"

The Caine Mutiny was one of the most popular fictions of the 1950s, appearing as a leading best-seller (1951), a courtroom drama on Broadway (1953), and a Hollywood film (1954). Briefly analyzed by Stephen Whitfield, Michael Rogin, and William Whyte, it is a good deal more familiar to historians than *White Christmas.* Still, it is odd that such a popular treatment of "mutiny," appearing as it does during the most repressive years of the Cold War, has not attracted more scholarly curiosity. While most readings regard it as a drama with serious political significance, usually overlooked is the therapeutic sentimentality at the heart of its conservative politics.[24]

The *Caine* is part of "the junkyard navy," a World War I–era destroyer rigged as a minesweeper for service during World War II. Used primarily for convoy and other undramatic assignments, only once does it actually sweep mines. "Ninety-nine percent of what we do is routine," explains the *Caine*'s caustic, shipboard novelist, Lt. Tom Keefer. We first see the ship through the eyes of young Ens. Willie Keith, a coddled, eager-to-please Princeton graduate who enlisted in the navy to avoid the army. In the novel, Keith hopes for a cushy staff position away from combat. The film, offered to the navy for script approval (voluntary censorship), avoids this minor embarrassment. In either case, Keith is shocked by the sorry state of the *Caine.* It's a "slack ship. The men act like a pack of cutthroats and the decks look like a Singapore junk." Everyone is desperate for reassignment, including the brusque, sarcastic Captain de

Vriess. The crew of "outcasts" is sloppy and irreverent, to be sure, but the ship performs its major assignments well, and the men are devoted to their tough commander. Loyalty is based on the fact that "all De Vriess cared about was results." His tyranny did not extend to petty regulations or military etiquette.[25]

When his reassignment finally comes through, De Vriess is replaced by his opposite, Philip Queeg. Where De Vriess was slack on military formality, Queeg is obsessed by it. Where De Vriess gave single-minded attention to big assignments, Queeg botches every one through cowardice, incompetence, or distraction. Where De Vriess's authority was backed by strength and skill, Queeg's tyranny grows out of weakness and insecurity. Queeg is nervous, compulsive, arbitrary, and increasingly paranoid. Under pressure he rolls steel balls in his hands. Brilliantly played by Humphrey Bogart in the film, he seems always on the brink of psychological collapse.

Queeg calls himself a "book man," but quickly makes clear that his own rules supersede all others."There are four ways of doing a thing aboard ship—the right way, the wrong way, the Navy way, and my way."[26] At first young Keith is attracted to Queeg's attention to detail—just what the sloppy ship needs. But the commander's obvious failings cannot be ignored. He runs over his own towline while ranting over an untucked shirt. Then he covers up his mistake by lying to higher command and blaming his men. Queeg also proves himself a coward. Whenever the ship enters a combat zone (once in the film, twice in the novel) he moves to the safer side of the ship and orders a premature withdrawal. The officers begin calling Queeg "old yellowstain."[27]

Cowardice, along with the irrational, draconian exercise of power (at one point he turns the ship inside out over some missing strawberries), leads the executive officer Steve Maryk to begin keeping a log of the captain's behavior. Lt. Keefer encourages Maryk to doubt the captain's sanity and tells him about Article 184 of *Navy Regulations,* which allows the second-in-command, in "extraordinary circumstances," to relieve the commander of his duties. When Maryk is prepared to report Queeg's erratic behavior to Admiral Halsey, Keefer loses nerve and Maryk refuses to make the case alone. The next crisis, a ship-threatening typhoon, triggers the "mutiny." Queeg's poor judgment threatens to swamp the badly listing ship. When he refuses to follow sound advice, Maryk invokes Article 184 and seizes command.

At this point, one can only feel relief. At last the ship is in safe hands and the dangerous captain at bay. In every respect this would seem to be what Stuart Hall calls the "preferred meaning" of the text.[28] Surely the actions of Maryk and Keith (who backed Maryk's "mutiny") were justified. It is a shock, therefore, to learn not only that they will be court-martialed, but that their acquittal is in real doubt. Their lawyer, Barney Greenwald, is reluctant to take their case and expresses disdain for the mutineers. Nonetheless, Greenwald wins an acquittal by exposing Queeg's psychological instability (under hard questioning, Queeg starts rolling those steel balls and making irrational claims).

Once again we are invited to believe that justice has been served. We join the officers as they gather to celebrate. The party has hardly begun when Greenwald arrives, already drunk, and attacks the men for their lack of loyalty. Queeg, he argues, was a victim, not a villain. The officers, not Queeg, were in the wrong for their infidelity to the war-weary captain.[29] Men like Queeg had protected the country through the hard years before the war while the other officers were home advancing their careers. Greenwald reminds the men that Queeg had once asked the officers for support. After his first cowardly retreat, Queeg gathered the officers and said: "Command is a lonely job. It isn't easy to make decisions. Sometimes the captain of a ship needs help. And by help I mean constructive loyalty."[30]

Greenwald thinks the officers dishonored themselves by failing to rally behind their commander:

> You turned him down. . . . He wasn't worthy of your loyalty. So you turned on him . . . you made up songs about him ["The Yellowstain Blues"]. If you'd given Queeg the loyalty he needed, do you suppose the whole issue would have come up in the typhoon? Do you think it would have been necessary for you to take over?

Maryk agrees that it probably would not. Keith also sees the light: "If that's true, then we were guilty."

> "Ah, Willie," Greenwald replies, "You're learning. You don't work with a captain because you like the way he parts his hair. You work with him because he's got the job or you're no good." By the end everyone seems convinced that Queeg was a victim of disloyalty and that Keefer—the caustic critic who first raised doubts about Queeg—had been the spineless (faintly left-wing) instigator, the true "author of the *Caine* Mutiny."

While most critics rightly emphasize the story's authoritarian implications, its justification of uncritical loyalty, several issues have not received sufficient attention.

First, the celebration of loyalty hinges on a therapeutic model of sentimental militarism. Both film and novel (in slightly different ways) suggest that Queeg's worst decisions might have been avoided, prevented, or at least covered up. He simply needed more encouragement from his officers. The point of *The Caine Mutiny* is not so much that military authority deserves blind loyalty, but that it may require something more complicated—therapeutic loyalty. After

Queeg's initial plea for "constructive loyalty," Keith says, "he almost made me feel sorry for him." Keefer jumps in: "Don't be so sentimental, Willie." But it is precisely that sentimentality that the story endorses.[31] The idea that authoritarians can be reformed, pacified, or at least circumvented is a long-standing sentimental convention (e.g., Charles Dickens). If all tyrants cannot be converted and taken into the family (as in Frank Capra's 1938 film *You Can't Take It with You*), at least their abuses can be absorbed or averted without directly challenging their power (as in Capra's *It's a Wonderful Life*, 1946).[32]

The sentimental militarism of *The Caine Mutiny* also rests on the shaky assumption that Queeg represents the most dangerous leadership the military can produce. Americans can safely obey all established military authority if the worst dictators they must endure are, like Queeg, more pathetic than ruthless. Queeg does finally endanger his men, but not by sending them into hopeless combat. Some men complain not that he is too tough, but that he is not tough enough—an "old lady."[33] The sentimental conceit is that the American military does not produce truly dangerous militarists—power-grabbing, war-loving, monsters—but that in times of crisis a few mediocrities may be promoted, petty tyrants who are afraid of their men and of their superiors.

The militarist politics of this story may be served even if viewers reject Greenwald's insistence that the officers had been wrongfully disloyal. One can support the mutiny against the incompetent Queeg without challenging military orthodoxy. After all, shouldn't a strong military get rid of its dead wood? The film opens with the disclaimer that no mutiny has ever occurred on an American naval vessel, but that offers viewers all the more reason to relish a fictional one that seems so amply justified. And the jaunty Max Steiner score sounds like it encourages the reluctant rebels. The fact that brave officers take control of a foundering ship may do much more for the navy's reputation than the implausible claim that subordinates should provide nurturing therapy to terrible commanders. Moreover, if we side with Maryk and Keith, we are allowed a flirtation with "mutiny" without the substance of real revolt. This is a supremely polite mutiny. It does not even require handcuffs, never mind a mob of angry sailors brandishing ropes and swords. Indeed, the crew is virtually absent from the story.

To be sure, as John Fiske argues, popular culture is comprised of many meanings ("polysemous" is, I think, the jargon).[34] Yet textual analysis suggests that cultural products may reject as many interpretations as they invite. *The Caine Mutiny* clearly rejects the legitimacy or threat of a genuine rank-and-file mutiny. For those in search of a subversive subtext, the best it has to offer is Article 184, a dignified restoration of legitimate power exercised by the next-in-command. The sailors grumble a bit, but show no sign of revolutionary potential.

The movie embraces only one rebellion unequivocally—the rebellion of Ensign Willie Keith against his domineering mother. Knowing that his mother would disapprove, the upper-class Keith keeps secret his relationship with a working-class, Italian-American nightclub singer. The mother also discourages any romance with the navy. "Don't volunteer or do anything foolish or dangerous."[35] The necessity of being *dis*loyal to his overbearing mother is as important to Keith's development as Greenwald's insistence that he learn to be more loyal to military authority. "Momism" must be destroyed to make way for militarism.[36] But, of course, Keith's fidelity to the navy is defined not as authoritarianism so much as democratic self-sacrifice. Keith rejects his mother's offer to pull strings to get him out of combat. Instead he takes the advice of his father, in a deathbed letter, who urges him to "be a man." Being a man means casting his lot with the unglamorous Caine, thus spurning his Princeton pretensions and his mother's connections. Keith's willingness to stay aboard the *Caine* dramatizes one of the principle features of sentimental militarism—the claim that the military is not so much an authoritarian class system as a democratic melting pot. The military is also credited with giving Keith the democratic strength to ignore his mother's prejudices and pursue his working-class girlfriend.

That said, the story's brief on behalf of military democracy is as strained and unconvincing as Greenwald's plea for therapeutic loyalty. A few minutes inside the prickly officer's mess on the *Caine* may well make one wish Keith had used his pull for a better assignment. The comradeship and diversity that were standard issue in Hollywood combat films of World War II is only a faint echo in *The Caine Mutiny.*

The Caine Mutiny suggests another, more speculative, thought. For audiences in the early 1950s the military figure most in the minds of Americans was almost certainly Douglas MacArthur (Eisenhower, like General Waverly, had already taken on a largely civilian identity). In crucial ways Queeg was no General MacArthur. While Queeg was a defensive lifer, fearful that his modest career would crumble with the next crisis, MacArthur was a five-star general who viewed war as a "romantic, mystical, and religious" experience.[37] Some parallels, however, may very well have resonated.

A considerable number of veterans, for example, hardly shared the view of MacArthur as a fearless warrior. Even before he withdrew from the Philippines during World War II, he acquired the nickname "Dougout Doug" for holing up on Corregidor while his troops were beleaguered on Bataan.[38] And during the 1948 presidential election, in which MacArthur allowed his name to be put forward in several primaries (while he was still in Japan as Supreme Commander for Allied Powers), "Veterans-Against-MacArthur" clubs formed in major cities to oppose his candidacy. Critics of MacArthur might also have seen parallels to Queeg in his vanity, arrogance, and unwillingness to admit his mistakes. Even William Manchester, his admiring biographer, concedes that MacArthur had paranoid tendencies. "Somebody else had blundered. MacArthur didn't make mistakes. Other men did, undermining him, making his tasks harder."[39]

Yet the outpouring of sympathy and support for MacArthur in April 1951 when he was relieved of command by President Truman was truly extraordinary. Tens of thousands of telegrams were sent to Washington in protest, cities held huge parades, and polls indicated that a strong majority opposed Truman's decision. Perhaps some viewers were more inclined to sympathize with Queeg given MacArthur's firing. The film suggests that removing commanding officers might be legal (i.e., Article 184 and the Constitution may sanction it), without being commendable.

With the Cold War, America was coming under new command and had somehow to justify a permanent, less glorious, military culture replete with the frustrations, uncertainties, and inconclusiveness of the Cold War stand-off. It is extremely significant that two of the period's most popular fictions—*The Caine Mutiny* and *Mister Roberts*—focus not on the high military drama of World War II but on events that occur in the "junkyard navy," far removed from the famous battles.

Part of the function of sentimental militarism in Cold War culture was to honor and legitimize the drudgery of maintaining a permanent, global, military establishment. As Herman Wouk honors the decommissioning of the *Caine,* he may also be saluting the commissioning of an enormous Cold War military to be deployed in endless, routine, and thankless operations:

> It's a broken-down obsolete ship. . . . It has no unit citation and it achieved nothing spectacular. . . . But we will remember the *Caine,* the old ship in which we helped to win the war. *Caine* duty is the kind of duty that counts. The high-powered stuff just sets the date and place of the victory won by the *Caines*.[40]

Mister Roberts and Democratic Militarism

Mister Roberts first appeared in the *Atlantic Monthly* (1946) as a series of loosely knit stories by Thomas Heggen. They were published as a novel the same year. Paperback editions circulated widely in the late 1940s and 1950s and in 1948, Heggen and Joshua Logan drew on the novel to develop an extremely successful Broadway show that ran for two years (1948–1950) before embarking on a nine-month national tour (1951).[41] In 1955 *Mister Roberts* took on yet another life as a John Ford/Mervyn LeRoy film starring Henry Fonda, who had played the lead in the stage version. A first audience poll of at least 15 million votes cast in 6,500 theaters named *Mister Roberts* the "best picture" among twenty choices for 1954–1955.[42]

Doug Roberts is a twenty-six-year-old naval lieutenant who dropped out of medical school after Pearl Harbor. According to the novel, Roberts had tried to enlist in the Abraham Lincoln Brigade during the Spanish Civil War but that war had ended before he could go. During the Cold War such volunteers were red-baited and black-listed as "premature antifascists." It is hardly surprising that this faint implication of left-wing political roots is dropped from the post–1946 play and film.[43] What is most strongly developed in all versions of *Mister Roberts* is his intense desire to get into combat before World War II ends. He avoids ideological explanations of his motives: "I just happen to believe in this thing" and "this is our last big push in the Pacific and I'm gonna catch it."

What Mr. Roberts catches instead is a supply ship called the *Reluctant,* so far from combat the shipboard doctor has never even seen a battleship.[44] The careerist, solipsistic Captain Morton (James Cagney) proudly maintains a palm tree on the bridge, a trophy from the admiral for "delivering more toothpaste and toilet paper than any other navy cargo ship in the safe area of the Pacific." Roberts is distraught. Every week he puts in for reassignment to a combat ship and every week the captain throws it away. Captain Morton hates Roberts because he is a "smart aleck college officer" who openly defies him (Morton grew up poor and uneducated, working his way up through the merchant marine). Despite the hostility, the captain knows that Roberts's competence improves his own chances for promotion.

The crew loathes Captain Morton. Their disdain is especially striking in the novel. When Morton asks one of the sailors what the men think of him he is told: "They think you're a prick."[45] Like the *Caine*'s Queeg, Captain Morton is arrogant, petty, self-centered, and insecure (he vomits when upset). If Barney Greenwald was livid about the failure of the *Caine*'s officers to buck up their unstable captain, what must he have thought of the insolent and insubordinate behavior aboard the *Reluctant*? He might have laughed. For it is soon evident that this story is as much a comedy as a drama. Though the genres sometimes blur, the laughter is rarely nervous and neither the captain's pettiness nor the crew's disloyalty seem capable of producing a serious disaster. Where Queeg's eccentricities are menacing and seem to cover a far deeper pathology, Morton's are mere bluster. He may be a fool, but so long as his tyranny is limited to an occasional tantrum and obsessive concern about his prize palm tree, neither his rule, nor the men's hatred of it, pose a serious threat to the ship's safety or hierarchy. And while the crew is sometimes able to circumvent Captain Morton, only their champion, Mr. Roberts, directly defies him. The men love Roberts for standing up to Morton. With awed expressions they say things like: "He'll fix the old man's clock—you wait!"

The men have not had liberty for a year and are beginning to "go Asiatic." When Mr. Roberts uses a bribe to get the crew shore leave in a "Polynesian paradise," the captain cancels it just as the ship pulls into port ("get them cannibals off this vessel"). Roberts is enraged. Storming to the bridge, he confronts the captain:

> How did you get in the Navy? How did you get on our side? Why you ignorant, arrogant, ambitious . . . keeping 62 men in prison. . . . How'd you ever get command of a ship? I realize in wartime they have to scrape the bottom of the barrel but where'd they ever scrape you up?

The captain proposes a deal. If Roberts will agree to stop requesting a transfer to combat duty and stop talking back in front of the crew, the men can have their liberty. Roberts agrees.

The crew goes on liberty and wreaks such havoc the local authorities order the ship to leave the port after a single night. Though the drunken and violent crew trashes the French colonial governor's home, hospitalizes thirty-eight army soldiers in a brawl, and rips the clothes off six island women, Roberts is charmed by the "crazy lugs."[46] He believes the night of liberty turned "sixty-two individuals" into "a crew." "In one night these guys are bound together." The captain, however, views the behavior as a blot on his record and is nastier than ever. A man of his word, Roberts honors his deal and refuses to oppose the captain as he had in the past. The crew turns on Roberts ("Do you really think he's buckin' for promotion?"). Roberts was never just one of the guys, but he had always been looked to as a surrogate commander who could be trusted to represent the best interests of all.

The men eventually discover that Mr. Roberts had been on their side all along, winning their liberty in exchange for loyalty to the captain. They respond by forging a letter to get Roberts transferred to a combat ship. Finally, in the combat he had longed for, Roberts is killed in a kamikaze attack. His final letter to his old shipmates becomes the film's benediction: "The unseen enemy of this war is boredom. . . . And I know now that the ones who refuse to surrender to it are the strongest of all; that at a time in the world when courage counted most, I lived among sixty-two brave men."

The same age that produced combat films like *To Hell and Back* (1955) moved mass audiences to tears with the claim that supply soldiers who battle boredom are "the strongest of all." This sentiment must have flattered the millions of World

War II veterans who never fired a weapon. Yet the film also resonates with postwar critiques of corporate tedium and alienation. Here was a story that made war seem even more boring than peacetime, an appealing tonic for those stuck in routine jobs with annoying bosses. It even adds the bolstering news that boring jobs are essential. This point was, in fact, a commonplace of American Cold War discourse. For every critique of corporate alienation, there were dozens of commencement addresses and rotary club speeches extolling the importance of civilian jobs to the national interest. In his first inaugural Eisenhower claimed that "each citizen plays an indispensable role" in the defense of freedom. "The men who mine coal and fire furnaces and balance ledgers . . . and heal the sick and plant corn—all serve as proudly, and as profitably, for America as the statesmen . . . and the legislators." Citizens "must be ready to dare all for our country," but the challenges of the Cold War were more likely to involve patient, undramatic, everyday service without a clear or decisive end in sight. Kennedy made national service more dramatic, but even he warned of a "long, twilight struggle."[47]

Roberts, the idealistic leader of the ship, redeems the virtue of military leadership by effectively displacing the decidedly cynical leadership at the top. Where Roberts is a model of patriotic self-sacrifice, the captain is, as Ensign Pulver put it, "an unpatriotic old slob." While Henry Fonda (age 51) is slightly unconvincing in the film portrayal of the youthful Roberts, the role is intended to represent a new breed of young officers coming of age during World War II. Unlike *The Caine Mutiny,* which offers Greenwald's homage to the old guard lifers (however incompetent), *Mister Roberts* is more a tribute to citizen soldiers and sailors who despise and circumvent careerist lifers like the captain. It might well be viewed as the first presidential campaign film on behalf of former naval Lt. John Kennedy (just as *White Christmas* was better suited to Eisenhower).

Without challenging the military establishment and its various missions, *Mister Roberts* preserves the sentimental idea that military service fosters democracy rather than authoritarianism. More than that, *Mister Roberts* links democratic antiauthoritarianism to manhood itself. Standing up to the captain becomes a test of manly patriotism and duty. When Mr. Roberts decides to throw the captain's beloved palm tree overboard, he marches to the task with military pivots and throws a crisp salute before carrying out the deed. And, to Ensign Pulver, who is always bragging about his fantastic plots against the captain but lacks the courage to follow through, Roberts says: "The day you pull a plot and admit it to the captain, that's the day I'll look up to you as a man." When Roberts leaves for combat, Pulver replaces him, yet still cowers before the captain. Upon news of Roberts death, Pulver fully replaces Roberts by destroying the captain's new palm tree.

Mister Roberts can be read as a tribute to antiauthoritarian individualism. Even the fact that most of the men do not share Roberts's courage might be taken as an implicit suggestion that the military as a whole has produced petty, careerist officers and numbed acquiescent, draft-induced enlisted men. Yet *Mister Roberts* offers only modestly subversive pleasures. Its antiauthoritarianism is largely founded on the kind of drunken, sexist, and sometimes violent male-bonding that many (today at least) would find more disturbing than humorous. There is also an elitist ring to Roberts's condemnation of Captain Morton ("Where'd they ever scrape you up?"). Nonetheless, *Mister Roberts*'s lampooning of the "Old Man," more than most Cold War popular narratives, calls into question the motives and reliability of military authority.[48] Comedies allowed a safer site for Cold War criticism.[49] In serious dramas, the "Old Man" authority is typically treated as a lonely, often isolated figure, who can be irascible and difficult but whose sacrifices are essential to the nation's well-being.

Reluctant Warriors, Lonely Bosses, and Understanding Wives

Though many popular films and novels of the 1950s are conditioned by Cold War concerns about loyalty, security, and authority, few deal directly with specific Cold War events. *The Bridges at Toko-ri* is an important exception. James Michener's novella was originally published in its entirety by *Life* magazine (July 6, 1953). With five million copies of this edition, *The Bridges at Toko-ri* was among the most widely read fictions of the year.[50] The 1954 film version, a faithful adaptation of Michener's story, was not a blockbuster, but it was certainly the most popular Korean War film of the 1950s and 1960s. *The Steel Helmet* (1951) and *Pork Chop Hill* (1959) may be better known by film historians but neither reached as wide an audience.[51] Unlike the war itself, *The Bridges at Toko-ri* provides a lean, uncomplicated plot. The hero, Harry Brubaker (William Holden), is a Denver lawyer and a bomber pilot in the unattached naval reserve. Since he had already fought in World War II, he is shocked and bitter to be reactivated for service in Korea. Nonetheless, he does his job bravely and skillfully. All the action culminates in a bombing run against the allegedly crucial bridges at Toko-ri.[52] The mission is successful, but Brubaker's fuel tank is hit by flak and he is forced to make an emergency landing in North Korea where he is killed by ground forces.

Like *The Caine Mutiny* and *Mister Roberts, The Bridges at Toko-ri* celebrates Americans whose service lacks a powerful sense of purpose. Bombing bridges is certainly more conventional heroic fare than anything offered in the "junkyard navy," but even Admiral Tarrant (Fredric March), commander of Brubaker's aircraft carrier, does not regard Korea as a decisive war. As he puts it in the novella, Americans should realize (but do not) that the nation is "engaged in an unending war of many generations against resolute foes who [are] determined to pull it down." The future would bring more "desultory battles" that would "stagger on" in places men would hate, but where they must fight if America were to meet the "honorable responsibilities forced upon her."[53]

Other than the combat scenes (which actually fill very little screen time) most of the dramatic tension centers on Brubaker's initial bitterness about the war in Korea. Quite possibly he expresses more explicit criticism of American Cold War policy than any hero of a popular American film from the early- to mid-1950s. If so, nothing could be more instructive of the narrow boundaries of mainstream opinion in this period. For Brubaker's dissent is more self-interested than principled and is effectively, if unconvincingly, transformed into consent. His objections come early in the film in a talk with Admiral Tarrant. The admiral invites the frank discussion because he is drawn to Brubaker, who reminds him of the two sons he lost in the Second World War. Brubaker complains: "I had to give up my home, my law practice, everything. Yes, I'm still bitter." (In the novella Brubaker also says that back home "nobody even knew there was a war except my wife. Nobody supports this war.")[54] When the admiral concedes that "militarily this war is a tragedy," Brubaker is emboldened to say: "I think we ought to pull out."

"Now that's rubbish, son, and you know it," the admiral rebuts. "If we did, they'd take Japan, Indochina, the Philippines. Where would you have us make our stand—the Mississippi?" The domino theory does not seem to sway Brubaker. One of the most remarkable things about the film is Brubaker's lack of convincing devotion to the war. We are invited to believe that he goes through a conversion from doubt and bitterness to conviction and noble self-sacrifice, but how and why remain a mystery. He fights on to his death, to be sure, but neither ideology nor patriotism explain his motives. Perhaps he's moved by a desire to demonstrate his courage (a good deal of attention is lavished on his struggle to overcome fear), or to keep from letting down his fellow pilots. But every explanation falls as flat as Brubaker's final cryptic remark. As he's being gunned to death by advancing Koreans, he announces: "I can see now [the admiral] was right. You fight simply because you're here."[55] This epiphany hardly provides an inspiring Cold War rallying cry.

At least Brubaker has someone shooting at him, itself a tried-and-true combat motivator. In *The Strategic Air Command* (1955), however, Jimmy Stewart is converted to Cold War orthodoxy with no enemy in sight. The enemy is not even named! Stewart plays Dutch Holland, an aging baseball star and World War II pilot who is pulled back into active duty by the Strategic Air Command—"the boys that drop the A-bomb." Like Brubaker, Holland is upset. "Where's the fire?" he asks. "I just don't see the necessity. . . . This is the craziest thing I've ever heard of . . . I've *done* my share." But Holland has no choice, and a few characters take turns trying to convince him and the audience that SAC is absolutely necessary—"the only thing that's keeping the peace" because "you never know when the other fella might start something." No wonder scholars have trouble defining the Cold War "other." He's merely the "other fella," at once colloquially familiar and mysteriously vague. Anthropologists from outer space would be as puzzled as Dutch Holland: "Where's the fire?"

Dutch is not alone in his confusion and resentment: "You sure have my sympathy, Dutch. I was yanked back in too. . . . I did my share once. We're not in a war." By mid-film, however, everyone has stopped complaining except Dutch's wife. What changes their minds? A major clue is provided by the repeated interpolation of long, loving shots of the B-36. The dialogue may be only so much babble alongside the doting footage of the shiny bomber, trailing six ribbons of puffy cotton against lurid sunsets and periwinkle skies, all to the tune of "The Air Force Takes Command." This may be "excruciatingly boring" to some viewers (my wife's comment on the film), but Dutch is utterly seduced and challenged by the latest military hardware ("you wouldn't believe the mess of new gadgets!"). When General Hawks, the tough, humorless SAC commander, gives Dutch a peek at the new B-47, the dimly lit hanger becomes half sanctuary, half brothel: "She's the most beautiful thing I've ever seen in my life, General. Just look at 'er, look at 'er. I sure would like to get my hands on one of these."[56]

No surprise, then, that Dutch decides to stay in the Air Force when his hitch is up instead of returning to baseball. His wife, Sally (June Allyson), is furious.[57] She's tried to be a "good sport" but she thought all the deprivations of being an "Air Force wife" would be temporary. And there's not even a war! Dutch had once had the same complaint but now he argues that there is "a *kind* of war." His only convincing motive (getting his hands on that B-47) is left unspoken and Sally's anger is hardly defused by Dutch's brief stab at Cold War rhetoric.

She takes her grievance to higher authority, General Hawks. She screams: "How much more do you want?" Hawks is unruffled. He simply says he has no choice. Miraculously, this seems to convert Sally. The next time we see her she is "ashamed" of her protest.[58]

General Hawks's uncanny ability to motivate reluctant warriors and subdue their dissenting wives is shared by many hard-bitten "Old Men" in Cold War film dramas of the fifties. Ensuring the loyalty of women could be especially problematic. In *Above and Beyond* (1952) and *The Atomic City* (1952), the wives of an atomic bomb pilot and a nuclear physicist criticize the obsessive "security," surveillance, anxiety, and paranoia of the new era. In *The Atomic City,* a son begins sentences with: "Mom, *if* I grow up I want to . . ." The husbands struggle to placate or silence their wives, but sometimes the "Old Man" has to step in to make sure the women are "on the same team." In *The Bridges at Toko-ri,* Admiral Tarrant says "war is no place for women" but he does not depend on this sentimental cliché to gain their allegiance. He believes women must understand war's danger. During a leave in Japan, Tarrant lectures Brubaker's wife (Grace Kelly). She's told that her husband almost died ditching his plane in the ocean and that he would soon have to fly against the heavily fortified bridges at Toko-ri. She must know the risks, or bad news may prove too much to bear. Tarrant's own wife had crumbled when her sons died. She became an alcoholic and "all that was gentle and loving about her slowly withered away."

Like the June Allyson character in *Strategic Air Command,* Mrs. Brubaker is instantly transformed by the Old Man's words. She reports the results to her husband: "I know what the Admiral was trying to tell me. I had to face those bridges too. Well, I have and I'll be alright now." Her allegiance to the American mission in Korea may be as unconvincing as her husband's, but Cold War sentimental militarism does not rely on ideology or even on plausible story lines; it relies on faith in male authority. It was a faith that could be strained to the breaking point but it is always, in the end, reaffirmed.

Female loyalty in fifties films is tested not only against the uncertainties of the Cold War, but draws as well on the legacy of World War II. *The Man in the Gray Flannel Suit,* as Emily Rosenberg has shown, has much to suggest not only about the cultural construction of corporate and suburban life in the 1950s, but also about the legacy and meaning of World War II.[59] Early on we hear Betsy Rath (Jennifer Jones) complain that her husband Tom (Gregory Peck) has not been the same since the war. He has become more cautious, more guarded, less ambitious. "You've lost your guts and all of a sudden I'm ashamed of you." Betsy wants Tom to be both more successful and more principled. Tom believes the two goals might be mutually exclusive and he reminds Betsy that she pushed him to get the new, more pressured job because she wanted more money and a new house. She insists that wasn't her primary objective. "The real idea was that I wanted you to go out and fight for something again like the fellow I married, not to turn into a cheap, slippery 'Yes Man.'"[60]

The last time Tom Rath had really fought for something was in World War II, and Betsy suspects the war transformed him into a cautious organization man. What she does not yet realize is that Tom did more than fight in Europe; he had an affair with an Italian woman and left her pregnant. Near the film's end Tom learns that his former lover and their son desperately need economic help. Tom wants to support them but believes he must gain Betsy's support for the plan. He expects her "understanding" will come if only she realizes how much violence and fear he experienced in the war. After one tortured night, Betsy mysteriously loses her anger. She returns to Tom and agrees to support his foreign family.[61]

Betsy offers viewers a fantasy of saintly forbearance and magnanimity (in the final line of the film Tom says he "worships" her), a welcome relief from her previously harsh, demanding, and selfish persona. Tom undergoes a less dramatic transformation from understated anxiety (Gregory Peck may be able to lose his mind, but never his cool) to calm and measured confidence. Early in the film we might speculate that he is suffering from a mild case of post-traumatic stress disorder (the film suggests that "mental health" may be the nation's most serious problem). However, we need to recall that Tom's flashbacks include not only the horror of war, but an idyllic affair with a beautiful, grateful woman who never once told him he had "lost his guts." So World War II is a site of fantasy and nostalgia as much as stress. Presumably Tom can enjoy the memory of the affair more easily now that his understanding wife has freed it of guilt.

Tom also resolves his problems at work.[62] He takes Betsy's advice and criticizes a speech drafted by the president of his company. Tom's honesty is appreciated and even though we know that several insufferable toadies have worked their way to the top of this organization, we are led to believe that Tom's integrity will work just as well. But how far does he want to climb? Like many other films of the period, *The Man in the Gray Flannel Suit* is concerned with the level of commitment America's mid-ranking officers should make to the corporation or the military. Should it be total or partial, permanent or temporary? Once again, the Old Man's advice is crucial. As in *The Bridges at Toko-ri,* the role is played by Fredric March. President of a major television network, the March character is haunted by his own failed family life—his separation from his wife and his rebellious, eloping daughter. "Don't let anything keep you away from your family!" he counsels Tom. Then a cloud descends over the Old Man's face and he unleashes a flood of bitterness:

> Somebody's got to do it. Somebody's got to dedicate himself to it. Big successful businesses just aren't built by men like you—nine-to-five and home and family. You live on 'em, but you never build one. Big successful businesses are built by men like me who give everything they've got to it—who live it body and soul, who lift it up regardless of anybody or anything else. And without men like me there wouldn't be any big successful businesses. My mistake was in being one of those men.[63]

Tom elects to remain a "nine-to-five" man, but the film is careful not to condemn the Old Men or the corporate world they rule. Grave, ambitious, unyielding, hard-working: these lonely men at the top often play small but crucial roles in films of the fifties. They may be feared, even disliked, but almost always they are respected as essential guarantors of the nation's military and economic well-being. Often they are not merely "lonely," but literally alone—estranged, divorced, or widowered, with children who have died or rebelled.[64] The junior officers and executives may not share the ruthless and isolating dedication of their bosses (opting for more stable family lives), but they faithfully serve their commander's mission. When personal doubts remain, the younger men find it reassuring to have strong, competent men at the top. Admiral Tarrant was nicknamed "George the Tyrant," but as James Michener writes: "the men who served with Tarrant soon forgot his tyranny and remembered his fantastic skill."

The heroes of many films of the 1950s reveal strains in Cold War militarism by raising doubts about the necessity and value of their service or by challenging petty, incompetent authority. Yet these very doubts provide a narrative means to reinforce military and corporate authority. By having doubters undergo conversions to commitment, viewers are invited to believe that their loyalty is founded not on fiat or mindless conformity but on thoughtful, individual, consent. Barbara Ehrenreich's analysis of the "gray flannel rebels" of 1950s popular fiction is helpful here. She argues that the male heroes in these novels are glibly aware of the danger that their jobs and lifestyles can turn people into conformist robots. By knowingly choosing such a path, however, they persuade themselves that they are not truly conformist. Rather, they have adopted a "higher conformity."[65] If some viewers find these conversions strained, forced, or unconvincing, sentimental conclusions have the power to erode much skepticism. By the end of *The Bridges at Toko-ri,* for example, we may well have forgotten Harry Brubaker's initial dissent ("I think we ought to pull out"). We are left instead with a line Ronald Reagan was still quoting in the 1980s, Admiral Tarrant's wistful "where do we get such men?"[66]

I have tried to suggest some of the ways popular fictions of the early Cold War construct the values and assumptions attending the permanent militarization of American society. These fictions reveal significant strains in their sometimes tortured efforts to convince reluctant Americans that their allotted roles are both freely chosen and "essential" to the national interest. Some stories, like *White Christmas,* deny the reality of militarization even as they support sentimental deference to military authority. Others, like *The Caine Mutiny,* make a brief on behalf of "constructive loyalty," founded on the sentimental faith that American commanders can be safely obeyed because faithful officers have the capacity (if not always the sufficient will) to uphold even the most unstable members of the military establishment. Others, like *Mister Roberts,* lend support to militarism by suggesting that military experience, however pointless or boring, is a bastion of democratic antiauthoritarianism that works to preserve rather than subvert American idealism. And films like *The Bridges at Toko-ri, Strategic Air Command,* and *The Man in the Gray Flannel Suit* support the militarization of Cold War culture by suggesting that reluctant military and corporate warriors, and their even more reluctant wives, may have their doubts about the postwar world but will do what they must to support the dedicated Old Men at the top. Their obedience is rewarded by the consolation of domestic harmony and affluence. The sentimentality is strained, but seductive.

Some of these stories may be read as Cold War variations of the biblical story of Abraham and Isaac in which Old Men offer up their sons to the service of the state. The fathers may acknowledge their torment in making such costly sacrifices, but these austere, silent men do not lose faith in the fundamental justice of the enterprise. The fifties was a time when a good many boys still referred to their fathers quite literally as "the Old Man." And, even the more approachable "dads" of the age had considerable authority over their children. Yet the strains revealed by these Hollywood fathers and sons expose a haunting foreshadowing of the generational tensions that would crack open in the 1960s. And unlike the bibilical story of Abraham and Isaac, God would not intervene to prevent the battlefield sacrifices of the Cold War. By the 1960s millions of young people discovered the cost of their faith in the jungles and villages of Vietnam and turned on the Old Men who ordered or supported the killing.[67]

Acknowledgments

I want to thank Peter Kuznick and James Gilbert for their astute editorial advice in preparing this chapter for publication.

Notes

1. These words are Irving Berlin's in the film *White Christmas* (1954).
2. Laurence Bergreen, *As Thousands Cheer: The Life of Irving Berlin* (New York: Viking, 1990), 409. "Around the world, GIs began inundating the Armed Forces Radio Service to play ['White Christmas']. In short order it became, quite spontaneously, the American soldier's anthem of longing and homesickness. . . . During its first ten years of existence [1942–1952] it sold three million copies of sheet music and fourteen million records." It would continue to sell well throughout the fifties and was recorded by dozens of artists including The Drifters (1956) and Elvis Presley (1957).

When Berlin heard Elvis's "rock-and-roll version" of "White Christmas" he was outraged. He had his staff plead with radio stations around the country to pull the song. A Portland, Oregon, station (KEX) actually fired a deejay for playing Presley's cover (whether they were acting at all in response to Berlin's pressure is unclear). A Los Angeles disk jockey (for KMPC) refused to play the song despite requests: "No, I won't play it," he explained. "That's like having Tempest Storm [the stripper] give Christmas gifts to my kids." See Arnold Shaw, *The Rockin' 50s* (New York: Da Capo Press, 1974), 157, and Bergreen, *As Thousands Cheer,* 513. On the importance of nostalgia and sentimentality in wartime swing music, see Lewis A. Erenberg, "Swing Goes to War: Glenn Miller and the Popular Music of World War II," in Lewis A. Erenberg and Susan E. Hirsch, eds., *The War in American Culture: Society and Consciousness during World War II* (Chicago: University of Chicago Press, 1996) 144–65.

3. On the "storylessness" of the Cold War, see Tom Engelhardt's superb book, *The End of Victory Culture* (Amherst, Mass.: University of Massachusetts Press, 1998).
4. In creating *This Is the Army,* Irving Berlin drew on his experience in World War I. In 1918 Berlin was drafted and sent to Camp Upton on Yaphank, Long Island, where the already famous songwriter organized an all-soldier revue called *Yip, Yip, Yaphank* that received as much attention as a hit Broadway show. See Bergreen, *As Thousands Cheer,* 150–73. General Eisenhower played a key role in sending the stage version of *This Is the Army* on an international tour of military bases. After seeing the revue in London on February 6, 1944, he made the request to General Marshall. The company then performed for the remainder of the war at dozens of bases, large and small, around the world. See Bergreen, *As Thousands Cheer,* 432–39. For excellent socio-political positioning of *This Is the Army,* see Lary May, "Movie Star Politics: The Screen Actor's Guild, Cultural Conversion, and the Hollywood Red Scare," in Lary May, ed., *Recasting America: Culture and Politics in the Age of Cold War* (Chicago: Chicago University Press, 1989), 136–38.
5. James Agee, *Agee on Film* (New York: McDowell, Obolensky, 1958), 53.
6. Allan Berube suggests that the drag numbers in *This in the Army* created a "subplot about homosexuality" (perceived at least among gay members of the cast and audience) that worked to subvert heterosexual orthodoxy. While it may be too much to claim any "plot" for this film, Berube's basic point is quite superbly argued and demonstrates one of the ways an apparently "straight" piece of propaganda can offer pleasures and meanings that are unintended, unwelcome, and potentially subversive; Allan Berube, *Coming Out under Fire* (New York: Plume, 1990), 69–80.
7. *Life* (October 11, 1954), 159. According to Ed Jablonski, *White Christmas* was a top film moneymaker of 1954, grossing $12,000,000; see *Irving Berlin: American Troubador* (New York: Henry Holt, 1999), 280.
8. *Newsweek* (October 11, 1954), 110; *Life* (October 11, 1954), 159; *Time* (October 25, 1954), 87; *Commonweal* (October 29, 1954), 94; *Saturday Review* (October 30, 1954), 29. *Life* mislabels the minstrel show dance song as "Abraham" (a song from *Holiday Inn* that was not used in *White Christmas*). It was really "Mandy." This mistake is

symptomatic of a careless critical tendency to exaggerate the resemblance between *Holiday Inn* and *White Christmas.* My own view is that *White Christmas* is quite different from *Holiday Inn* and is at least as interesting as a historical text.

9. For the rise of the national security state and the Cold War, see Michael S. Sherry, *In the Shadow of War* (New Haven: Yale University Press, 1995); Walter LaFeber, *America, Russia, and the Cold War, 1945–1992* (New York: McGraw-Hill, 1993); Melvyn Leffler, *A Preponderance of Power: National Security, the Truman Administration, and the Cold War* (Palo Alto: Stanford University Press, 1992); and Thomas McCormick, *America's Half-Century: United States Foreign Policy in the Cold War* (Baltimore: Johns Hopkins University Press, 1989).
10. However, some of the moderately liberal films of 1947–1949 were box office successes: *Crossfire, Gentleman's Agreement, Pinky,* and *Home of the Brave.* Also, as Nora Sayre points out, while the most rabidly anticommunist films of 1947 to 1954 were not hits, they were screened as reruns on double features and were thus more widely viewed than earnings would indicate; Nora Sayre, *Running Time: Films of the Cold War* (New York: Dial Press, 1982), 40–48, 79–80.
11. Michael Rogin, *Ronald Reagan, the Movie* (Berkeley: University of California Press, 1987), 262.
12. See Terry Christensen, *Reel Politics* (London: Basil Blackwell, 1987), 74, 85 on audience drop. Tom Doherty, *Teenagers & Teenpics: The Juvenilization of American Movies in the 1950s* (Boston: Unwin Hyman, 1988), 1. As Doherty argues, the film industry in the fifties began increasingly to turn toward segmented markets, especially young audiences.
13. See, for example, John Fiske, *Understanding Popular Culture* (Boston: Unwin Hyman, 1989).
14. See Engelhardt, *The End of Victory Culture.*
15. While musicals are beginning to receive more attention by cultural historians, they surely merit much closer examination. They were one of the most popular film genres from the 1930s to the 1950s and might well be the basis for a cultural study every bit as concerned with mythology, ideology, and politics as Richard Slotkin's *Gunfighter Nation* (New York: HarperCollins, 1992). For suggestive possibilities on musicals in this period, see Andrew Dowdy, *The Films of the Fifties* (New York: Morrow, 1973); Ethan Mordden, *The Hollywood Musical* (New York: St. Martin's Press, 1981), 165–94; Rick Altman, ed., *Genre: The Musical: A Reader* (London: Routledge & Kegan Paul, 1981).
16. Irving Berlin, "What Can You Do with a General?" This song sounds like it was written merely to cover the plot line. It was written in 1948, however, and is in keeping with Berlin's enormous reverence for military commanders; Jablonski, *Irving Berlin,* 359.
17. Gen. Lucius Clay chaired the board of Continental Can, Adm. Ben Moreel was chair of Jones & Laughlin Steel, Gen. Omar Bradley was chair of Bulova Research Laboratories, Gen. James Doolittle was vice-president of Shell Oil, Gen. Albert Wedemeyer

was vice-president of AVCO Corporation, and Gen. Leslie Groves was vice-president for research at Remington Rand where Gen. Douglas MacArthur would soon become chair; C. Wright Mills, *The Power Elite* (New York: Oxford University Press, 1956), 214. Retired officers also found ample employment opportunities in the civilian agencies of government. See Sidney Lens, *Permanent War: The Militarization of America* (New York: Schocken Books, 1987), 15.

18. This lapse in the general's resolute acceptance of civilian status was manipulated by his loyal housekeeper who sent all his civvies to the cleaners.
19. Even "They Like Ike" went through a revision (in January 1951) after Eisenhower indicated some willingness to run for president. The original song features two Democratic Senators baiting a Republican about Eisenhower's reluctance to run. During this period Berlin told a reporter, "To tell the truth, I *love* Eisenhower. . . . This isn't a campaign song, but if anybody wants to use it as a campaign song, it's okay with me"; *New Yorker* (January 19, 1951), 21. Berlin's willingness to share the song is remarkable for a man so protective of his copyrights. In 1955 Eisenhower presented Berlin with a special medal "in recognition of his services in composing many popular songs, including 'God Bless America.'" Bergreen, *As Thousands Cheer,* 507–10. Edwin Diamond and Stephen Bates, *The Spot: The Rise of Political Advertising on Television* (Cambridge, Mass.: MIT Press, 1988), 61. That Eisenhower's popularity drew in part on a mood of deference is suggested by Michael Rogin: "Eisenhower politics was the politics of deference to responsible leadership" (Rogin, *The Intellectuals and McCarthy,* 247).
20. The eight hundred golf rounds comes from Curt Sampson who adds: "That's almost twice a week. Pretty good considering he had a heart attack in 1955, abdominal surgery in 1956, and a stroke in 1957"; Curt Sampson, *The Eternal Summer: Palmer, Nicklaus, and Hogan in 1960, Golf's Golden Year* (Dallas: Taylor Publishing, 1992), 5; Stephen Ambrose, *Eisenhower, The President* (New York: Simon and Schuster, 1984), 72–75.
21. Engelhardt, *The End of Victory Culture,* 86–89.
22. In *White Christmas,* an African American appears only once, briefly, as a train porter. It also includes a minstrel number "Mr. Bones" in honor of "the minstrel days we miss." This sentimental tribute to racist stereotyping was apparently deemed inoffensive since it was not performed in blackface.
23. In the run-up to the Army-McCarthy hearings of 1954, McCarthy attacked Gen. Ralph Zwicker for failing to block a routine promotion of a left-wing army dentist named Irving Peress. McCarthy claimed Zwicker, a hero of the Battle of the Bulge, was "a disgrace to the uniform." See Richard H. Rovere, *Senator Joe McCarthy* (New York: Harper Torchbooks, 1959), 30; or John G. Adams, *Without Precedent* (New York: Norton, 1983), 124–29.
24. The novel, by Herman Wouk (Garden City, N.Y.: Doubleday, 1951), placed second among fictional best-sellers for 1951, just behind James Jones's *From Here to Eternity.* Remarkably, it was also number two the following year. It sold 425,000 in hardcover

within two years. It did even better in paperback where, by 1975, it had a total sales figure of 2,087,173 copies. See Alice Payne Hackett and James Henry Burke, *80 Years of Best Sellers, 1895–1975* (New York: R. R. Bowker Company, 1977), 19, 155–59; Stephen J. Whitfield, *The Culture of the Cold War* (Baltimore: Johns Hopkins University Press, 1991), 60–63; Michael Rogin, *Ronald Reagan, the Movie,* 256–57. Rogin's analysis is suggestive but must be consulted with caution because he conflates the characters of Willie Keith and Steve Maryk; William H. Whyte Jr. *The Organization Man* (New York: Simon & Schuster, 1956), 243–48. A 1988 version of the 1953 play was made for television by director Robert Altman and can be viewed on video as *The Caine Mutiny Court-Martial.* The film is also available on video. It was made by Columbia Pictures and directed by Edward Dmytryk, one of the Hollywood Ten who, after serving part of his prison sentence for contempt of Congress, apologized for his prior resistance to the House Un-American Activities Committee and went before them to name names.

25. Wouk, *The Caine Mutiny,* 140, for the "results" quotation. "Slack ship" is from the film. "Ninety-nine percent" is in both novel and film.
26. This line appears in both novel and film.
27. The enlisted men may have similar feelings but this story focuses almost exclusively on officers. So, too, do *Mister Roberts* and *The Bridges at Toko-Ri.* My untested impression is that by the 1950s most war films focus on officers, and increasingly on naval and air force experiences. War films of the 1940s often featured army enlisted men.
28. Stuart Hall, et al., *Culture, Media, Language: Working Papers in Cultural Studies* (London, 1980), 117–38.
29. Greenwald's attack is far more incisive and powerful in the play (performed by Henry Fonda) and film versions (wonderfully played by Jose Ferrer). During some Broadway performances Fonda would give such a stirring banquet monologue (where Queeg is defended and the officers criticized) that he ended up in tears. According to Herman Wouk, the effect on audiences was "explosive"; see Howard Teichmann, *Fonda: My Life* (New York: New American Library, 1981), 228. In the novel, however, the drunken speech is weakened by Greenwald's rambling, incoherent suggestion that men like Queeg had prevented his Jewish mother from being turned into a bar of soap. In scenes like this, and in other odd ways, Wouk invites readers to feel superior to bigots even as he preserves anti-Semitic stereotypes. The same can be said of his treatment of black characters, such as the black steward's mates ("mess boys"), one of whom is named "Rasselas."
30. This explicit plea is missing from the novel. In every respect the novel's critique of the mutiny is less persuasive. Queeg is less sympathetic and Greenwald is not nearly so persuasive in his final attack on the men. However, according to William Whyte, book reviewers tended to applaud the defense of Queeg, while movie reviewers were a bit more skeptical. Whyte suggests that "the three-year time lapse between the book and the screen and stage versions gave them time for a double take."
31. At the end of his novel, Wouk shows the kind of "constructive loyalty" he has in

mind. Keith is now executive officer to the gutless Captain Keefer. After a kamikaze attack, Keefer jumps overboard in fright. Keith takes over but not before getting permission from Keefer. And then, after putting out all the fires, he rescues Keefer and covers up his embarrassing cowardice and neglect of duty.

32. Another example of authority tamed by sentiment comes in William Wellman's 1950 film *The Next Voice You Hear.* After God comes on the radio to tell the world to count their blessings and do their homework, the beleagured working-class hero draws kindness from his previously insufferable boss simply by deciding not to fear him.
33. Wouk, *The Caine Mutiny,* 173. Like James Dean's father in *Rebel Without a Cause,* Queeg invites mutiny because he fails to provide the strong, unequivocal, fatherly guidance the crew (and James Dean) would welcome. It is worth noting that the *New York Times* critic disliked the "thin theatrics" of the ending, implying that it was too feminine. "The body of the picture—the good, solid, masculine core—that has to do with the chafing of naval officers under a neurotic captain's command, is exciting and revealing"; *New York Times* (June 2, 1954), 17.
34. Fiske, *Understanding Popular Culture.*
35. Compare this to the mother in *Rio Grande* who arrives at her son's camp in an effort to buy him out of combat. See Richard Slotkin's insightful analysis in *Gunfighter Nation,* 464–67.
36. Rogin, "Kiss Me Deadly: Communism, Motherhood, and Cold War Movies," in *Ronald Reagan, the Movie,* 236–71.
37. William Manchester, *American Caesar* (Boston: Little, Brown, 1978), 629.
38. "During his seventy-seven days on Corregidor he visited the peninsula just once." Bataan was five minutes away by torpedo boat; Manchester, *American Caesar,* 235.
39. Manchester, *American Caesar,* 524; Joseph C. Golden, *The Best Years, 1945–1950* (New York: Atheneum, 1976), 376.
40. Wouk, *The Caine Mutiny,* 488.
41. Heggen's original novel was not a top ten best-seller in hardcover but, spurred by the play and film, it did become one of the top paperback best-sellers of the late 1940s and 1950s with a total of 2,246,396 copies sold; Hackett and Burke, *80 Years of Best Sellers,* 18, 41. Irving Berlin, incidentally, was an investor in the Broadway production; Bergreen, *As Thousands Cheer,* 479. In 1949, with the play at its height of success, Heggen committed suicide. He was only 29.
42. *New York Times* (December 7, 1955), 47: "The poll marks the first time that the public has had the opportunity, on a national basis, to vote year-end movie preferences." The film also enjoyed critical success. It was voted best feature film of 1955 by a poll of 354 film critics; *New York Times* (5 March 1956).
43. Thomas Heggen, *Mister Roberts* (Cambridge, Mass.: Houghton Mifflin, 1946), 161.
44. The name *Reluctant* is dropped in play and film and replaced with AK601.
45. Heggen, *Mister Roberts,* 55. This line and the nickname "Stupid" are cut from the play and film. The play and movie also cut out a line about "the clap" though Henry

Fonda occasionally put it back in when friends were in the audience; Howard Teichmann, *Fonda: My Life,* 184, 191–92. The only negative contemporary review of the Broadway play I have seen was written by Joseph Wood Krutch in the *Nation.* He found it "crude, simple, and completely adolescent," and compared the plots against the captain to schoolboy pranks against a nasty headmaster. "Stunned by its naivete," he was almost as shocked by the overwhelmingly favorable response it received from metropolitan audiences; *Nation* (April 10, 1948), 402. More typical was *The New Yorker* review—"the best new play . . . this season" (February 28, 1948), 46, and the *New Republic* —"one of the funniest plays ever seen on the American stage" (March 8, 1948), 29.

46. Acording to Joseph Wood Krutch, one of the two biggest laughs from a Broadway audience came when the drunken sailors were returned to the ship in a cargo net. The other came when Ensign Pulver exploded his firecracker; *Nation* (April 10, 1948), 402. The *New York Times* found the film's shore leave scene "screamingly funny"; *New York Times* (July 15, 1955), 14.
47. *Inaugural Addresses of the Presidents of the United States* (Washington, D.C.: Government Printing Office, 1989), 299, 295, 308. My comments are also based on reading commencement addresses of the period given at Notre Dame, Berkeley, and Amherst College.
48. A fruitful comparison may be made between *Mister Roberts* and *From Here to Eternity* (novel, 1951; film, 1953). In *From Here to Eternity* the martyr is not an idealistic officer but a working-class enlisted man: Robert E. Lee Prewitt. Yet the enlisted men heroes, Prewitt and Sergeant Warden, like the crew of the *Reluctant,* each struggle to assert some individuality within and against the authoritarianism of military life (Prewitt by refusing to box, and Warden through his affair with the captain's wife—see page 107 in the novel). In this instance, as well, antiauthoritarianism does not lead to a rejection of military life. Quite the reverse. Both Prewitt and Warden reject marriage to continue their lives in the institution that has given them the only secure and satisfying identity they have known. Male bonding (notice especially the film's drunken scene in which Warden gently strokes the hair on the back of Prewitt's head) stands as an antidote both to the military's brutality and to female demands for upward mobility. Warden's epitaph at Prewitt's death is: "He loved the army more than any soldier I ever knew."
49. In this context we need further studies of such figures as political cartoonist Herb Block, Walt Kelly ("Pogo"), comedians "Bob and Ray" and Mort Sahl, as well as film and television comedies such as *Sgt. Bilko, McHale's Navy,* and the Jerry Lewis army films.
50. A. Grove Day, *James A. Michener* (Boston: Twayne Publishers, 1977), 75. The *Life* edition preceded the official publication of the hardbound copy by three days. The book was not among the top ten best-sellers in fiction for 1953. The magazine edition of five million was far greater, however, than the publishing totals of the hardcover best-sellers. The top fiction seller for 1953 *(The Robe)* sold 188,000 copies; Hackett

and Burke, *80 Years of Best Sellers, 1895–1975,* 161.

51. *MASH* (1970) is surely the most widely viewed Korean War film though it is more a response to the Vietnam War than the Korean. Popularity estimates are drawn from the Blue Banner Movie Guide, Version 1.0, a software source on 9860 film titles published by Blue Banner Software, 1992.
52. I have not yet determined if Toko-ri was an actual target of American bombs (Michener and the film make no disclaimers), but if it was, its strategic importance is vastly exaggerated in this story. *Pork Chop Hill* also greatly exaggerates the significance of that battle to the peace talks.
53. James A. Michener, *The Bridges at Toko-ri* (New York: Random House, 1953), 42–43.
54. Michener, *The Bridges at Toko-ri,* 36.
55. In *Twelve O'Clock High* (1949), Gregory Peck has a far harder time overcoming the doubts and objections of the pilots in his B-29 wing. Finally his example, along with one crucially successful mission, seem to do the trick. But given the low morale at the outset, the radical turnaround remains one of this important film's mysteries. For example, at one point a young Medal of Honor winner confronts Peck with the following speech: "I can't see what good we're doing with our bombing and all the boys getting killed. . . . It's like we were some kind of guinea pigs. Only we're not proving anything . . . . It just doesn't make any sense. I just want out. I don't want to fly again. I want to transfer to another branch . . . . I just don't have confidence in anything anymore."

    Peck would have a similar difficulty in *Pork Chop Hill* (1959). He is confronted by a black soldier (Franklin) who refuses to fight and threatens to kill him: "I don't want to die for Korea. What do I care about this stinking hill. You ought to see where I live back home. . . . I ain't sure I'd die for that. It's a cinch I ain't gonna die for Korea." Peck produces another amazing Cold War conversion by convincing Franklin to fight by telling him that other guys in the company also hate Korea, that they're all probably going to die anyway, and that they might as well go down together. "Its a pretty exclusive club but you can still join up if you want to." In this film democratic bonding and inclusiveness redeem an otherwise meaningless mission. In the end, the film claims the last stand on Pork Chop saved the peace process and the freedom of millions, but the bulk of the film makes this framing unpersuasive.
56. H. Bruce Franklin, *War Stars: The Superweapon and the American Imagination* (New York: Oxford University Press, 1988), 186.
57. Though June Allyson's reputation is largely founded on roles in which she is implausibly sweet, perky, and deferential, in *The Shrike* (1955) her criticism and pressure drive her husband into a nervous breakdown.
58. Though Sally's capitulation is necessary to the film's politics—now everyone is loyal to the Strategic Air Command—she is spared the necessity of remaining an Air Force wife. It turns out that Dutch's arm was injured so badly in a crash landing that he can no longer fly or play baseball. He will return to civilian life and coach.
59. Emily S. Rosenberg, "'Foreign Affairs' after World War II: Connecting Sexual and

International Politics," *Diplomatic History* (Spring 1993), 59–70. The film is based on Sloan Wilson's novel, which was the number five best-seller in fiction for 1955; Hackett and Burke, *80 Years of Best Sellers*, 166.

60. In *Executive Suite* (1954), the June Allyson character voices a similar reaction: "I'll fight anything, anyone, even you, if I think it'll make you into something I can't go on loving."
61. In the novel, the marital reunion is not the only happy ending. The couple can look forward to a highly profitable real estate venture.
62. This resolution actually precedes his confession of the affair.
63. *New York Times* film critic Bosley Crowther described this scene as "one of the most eloquent and touching" he had ever seen; *New York Times* (April 13, 1956), 21.
64. The CEO in *The Man in the Gray Flannel Suit* and Admiral Tarrant in *The Bridges at Toko-ri* were both separated from their wives and lost sons in World War II. They also had a rebellious daughter or promiscuous daughter-in-law.
65. Barbara Ehrenreich, *The Hearts of Men* (Garden City, N.Y.: Anchor Books, 1984), 29–41.
66. Michael Rogin, *Ronald Reagan, The Movie*, 7. *Sixty Minutes* has traced the process by which Reagan first credited the line "Where do we find [*sic*] such men?" to the movie admiral in *Bridges at Toko-Ri*, then assigned that line to a real admiral, and finally quoted it as if he had thought of it himself.
67. I am indebted to James Gilbert for suggesting the basic idea of this paragraph.

CHAPTER 5

Sex, Gender, and the Cold War Language of Reform

Joanne Meyerowitz

For more than a decade now, American historians have posited links between Cold War ideology and post–World War II conservative gender and sexual ideals. Various scholars have argued that the Cold War assault on communism reinforced the subordination of women and the suppression of sexuality. In the most common variant of this argument, the fear of communism pushed middle-class Americans to look to masculine strength and the patriarchal home as protective forces in a dangerous world. In this anxious context, independent women, gay men, lesbians, "domineering moms," and "matriarchs," among others, seemed to threaten masculinity, the nuclear family, and the nation, just as communism seemed to threaten the international order. The Cold War containment of communism had domestic corollaries in the containment of women within the home and the containment of sex within heterosexual marriage. In her 1988 book *Homeward Bound,* Elaine Tyler May offers the most detailed and influential account of this strange marriage of foreign policy and domestic ideals, but Michael Rogin, John D'Emilio, Geoffrey Smith, Emily Rosenberg, and others also embrace it in various permutations. In scholarly essays, historical monographs, and popular histories, the anxieties of a Cold War culture now help explain the failures of the feminist movement, misogynist attacks on women, and homophobic assaults on gay men and lesbians.[1]

In this chapter, I do not attempt to sever the historiographic links between

Cold War policy and domestic ideology; rather, I hope to complicate and multiply them. While in many cases Cold War thought did indeed reinforce traditional gender roles and the heterosexual marital norm, in other notable cases it also seemed to subvert them. Certain reformers developed a distinctive Cold War language to make demands for gender equity and sexual freedom. This language of reform suggests that Cold War foreign policy had no fixed association with gender and sexual norms but instead had multiple and contradictory meanings. As Susan Hartmann has written recently, "conflicting elements" in Cold War thought not only helped "sustain cultural conservatism"; they also "promoted gender role changes."[2]

Without a doubt the Cold War had damaging political, social, and cultural ramifications: anticommunist politics decimated the American left, narrowed political thought, encroached on freedom of speech, and violated civil liberty. Nonetheless, as a political discourse adopted by a wide range of conservatives, liberals, and even radicals, Cold War language had different meanings and different uses in different political contexts. For certain reformers, the Cold War offered a new vocabulary for extending democratic ideals and demanding individual freedom.

This chapter focuses on the public Cold War language of two social movements of the 1950s: the women's movement and the movement for sexual freedom. It looks primarily but not exclusively at the magazine *Independent Woman,* which served as the official periodical of the National Federation of Business and Professional Women's Clubs,[3] and two magazines of the fledgling homophile (or gay rights) movement, the *Mattachine Review* and *ONE,* both of which pioneered in the public defense of homosexuality. All of these publications adopted Cold War language but did not find that language at odds with either gender or sexual reform. Instead they used the Cold War to promote liberal social change.

In the 1950s the organized women's movement was, as Leila Rupp and Verta Taylor have argued, "in the doldrums." Since the suffrage movement of the early twentieth century, feminism had lost much of its energy and most of its clout. The acrimonious internal divisions over the Equal Rights Amendment (ERA) and a long-standing public hostility to "militant" feminism reasserted themselves after World War II.[4] The bitter battles of the Cold War further contributed to the decline of the organized movement. As in the 1920s, some women's groups and some women reformers faced accusations of left-wing or "un-American" politics.[5] Internal red-baiting, for example, tore the Woman's International League for Peace and Freedom asunder, and the United States government hounded women on the left. The Congress of American Women, a left feminist organization of the late 1940s, disbanded after the Justice Department ordered its board to register as "foreign agents," and various women in the Communist Party faced harassment, arrest, and deportation.[6] In this climate the women's movement, like the labor movement and the civil rights movement, lost its openly left participants. Even women's groups with little or no left membership engaged in a kind of restrictive self-censorship.[7] In a seeming attempt to forestall any possible accusations of communist influence, some women's organizations adopted "only the most cautious" approaches to controversial issues.[8]

Still, the women's movement was far from moribund in the postwar era. As several historians have noted recently, women in the National Woman's Party, the Young Women's Christian Association, the Women's Bureau, and the labor movement, among others, continued to push for gender equity in law, politics, and employment.[9] The list should also include the National Federation of Business and Professional Women's Clubs (BPW), one of the larger, more active, and more openly feminist women's organizations of the postwar era.

Founded in 1919 to promote the interests of white-collar working women, the BPW, a nonpartisan, white, middle-class organization, espoused a liberal feminism that demanded equal access to careers.[10] By the post–World War II era, the BPW actively supported the ERA, equal pay legislation, and the appointment and election of women to government office. Historians usually associate this "equal-rights" feminism with the tiny National Woman's Party (NWP), but the BPW rivaled the NWP in its leadership of the ERA campaign and far outstripped the NWP in its membership.[11] In the early 1950s, the BPW had 160,000 members while the NWP had only 5,500.[12]

Like many other women's organizations, the BPW faced occasional red-baiting in the post–World War II era. In the late 1940s two of its leaders, Lena Madesin Phillips and Sarah Hughes, found themselves branded as left-leaning. And in the late 1950s, the right-wing magazine *American Mercury* revived the accusations when it listed the BPW among "women's organizations . . . notoriously infiltrated with socialistic and international activists." In this latter case, the BPW and its lawyers successfully demanded that the magazine recant. In 1959 the *Mercury* admitted that "the federation's strong anticommunist stand is well-known."[13]

The BPW had worked strenuously throughout the postwar era to establish its anticommunist credentials. Like most other mainstream women's organizations of this era, the BPW not only defended women's rights; it also staunchly denounced communism.[14] From the late 1940s, *Independent Woman*, the magazine of the BPW, published articles condemning communism and urging women to join the fight against it.[15] In 1950, when the United States entered the Korean War, the BPW "immediately pledged its support," and its leaders soon participated in various forums, conferences, and programs "to strengthen the free world against the dangers of communism."[16] In 1953 the federation's board of directors "unanimously adopted" a Declaration of Principle that recognized "the dangers which threaten our democracy from within and without" and asked members "to organize their programs and efforts to combat communism and all other forms of totalitarian propaganda."[17] A few years later the BPW advertised its pamphlet, "Women's Role in Combatting Communism."[18]

For the leaders of the BPW, though, support for the Cold War did not entail support for the containment of women within the home. Instead, the BPW explicitly used the Cold War to promote women's participation in the public realm. Deploying arguments articulated during World War II, the federation linked enhanced national security to women's greater participation in government and business. It repeatedly claimed that women constituted a key resource that the nation could not afford to waste. During the Korean War, for example, BPW leader Margaret A. Hickey urged members to "demand that women be recognized as full partners in all phases of our civil defense as well as in the military and economic mobilization." For Hickey, women constituted "the greatest unused potential for leadership in the world today."[19] As the nation mobilized for limited war, the BPW asked women to take advantage of the new "opportunities" available. "The need is here," one article announced, "for trained and skilled women."[20] The federation also endorsed registering women for the draft and helped recruit women for the armed forces.[21]

As the Korean conflict faded from the public eye, the BPW called for "greater opportunities" for women to participate in programs "for counteracting communist propaganda" and for more women in civil defense programs, United Nations delegations, and technical assistance missions.[22] In 1957, after the Soviet Union launched the Sputnik satellite, the BPW turned its attention to "science, mathematics, education, and teaching." Once again federation leaders suggested using "fully all the resources of woman power."[23] The underlying argument was clear: Cold War competition demanded that employers, educators, and government officials develop and use the talents of women. Cold War ideals converged conveniently with the BPW's long-standing commitment to using "women's full capacities in government and industry."[24]

The Cold War also proved useful in the BPW's efforts on behalf of the ERA. Just as African American civil rights leaders pointed to the racism that undermined American democracy and embarrassed the United States abroad, so BPW leaders pointed to the gender-based legal inequities that belied America's claims to equal rights for all.[25] As one article claimed, "Other countries looking to the United States for guidance, or desiring to find weaknesses on which to base propaganda and criticism, have made much of the legal situation of women in this great free country." For the BPW, the solution was simple: American women needed "the dignity of full citizenship under the Constitution."[26] In 1956 Hazel Palmer, BPW national president, stated the argument bluntly: "The prestige of the United States is assailed by Russia on the grounds that it does not grant its women equal legal rights. If our Constitution contained the Equal Rights Amendment, the United States would not be placed in the position of endorsing this principle in theory but being unable to put it into effect."[27] In this line of argument, the winning of the Cold War did not depend on maintaining domesticity; it relied on promoting equal rights.

The BPW was a prominent organization and an insistent one, but it was not the only women's group to use the Cold War to argue for reform. From the late 1940s on, for example, the American Association of University Women (AAUW) called for "full participation and contribution of women to the nation's defense." In 1948, it asked President Truman to establish a commission "to consider the integration of women into all phases of preparedness activities, including policy-making." And in 1950, it wrote to the Secretary of Defense asking for "removal of the statutory limitation restricting the strength of the women's components to two percent of the armed forces."[28] Throughout the 1950s, the AAUW *Journal* repeatedly promoted new careers for women by presenting women as a resource for the national defense. One article reported the "dangerous shortage of scientific manpower" as "a critical national problem," and then called for more women in the physical sciences. "The future of our national security and economic development," another article claimed, "depends upon woman power to a greater degree than ever before."[29] To this end, the AAUW called for better technical training, job guidance, and higher education for women.

In general the BPW and the AAUW portrayed women as potentially equal participants in the nation's defense, as "womanpower" that could remedy shortages in skilled and technical labor as well as in policy-making personnel. But occasionally the organizations also envisioned women as making different contributions than men. They occasionally implied that women as mothers and homemakers had special qualities needed for the nation's defense and international peace. Women, one article in *Independent Woman* pronounced, could make "a unique contribution" because they were concerned with the impact "of national and international politics on the life and homes of people." Because of women's special concerns, they could "humanize the vast issues daily presented to the public."[30] Similarly, an article in the *Journal* of the AAUW urged "women to participate in public life" because women's "particular insight and understanding" were "badly needed by the nation and the world in this chaotic, atomic age." Women in government could "help to bring peace and prevent annihilation of the human race."[31] While reformers used this maternalist construction of women as especially peace-loving and nurturant less frequently than they had in the early twentieth century, they came back to it occasionally to suggest that women's public participation would not only augment but also balance the nation's defense.

At the height of the Red Scare, several women's groups, including the BPW and the AAUW, joined together to use the Cold War as an opportunity for collective action on behalf of women. In 1950 the BPW called together representatives from national women's organizations to plan for "full partnership of women" in the nation's defense. The first women's conference met in October with representatives from thirty-eight organizations present. The women established an umbrella organization, a clearinghouse "to mobilize and utilize womanpower" for national security needs.[32] With some shifts in membership, the group organized in 1951 as the Assembly of Women's Organizations for National Security. In 1954 member organizations included, among others, the BPW, AAUW, General Federation of Women's Clubs, National Association of Negro Business and Professional Women's Clubs, National Association of Women Lawyers, Women's Division of the Democratic National Committee, and National Federation of Republican Women. Although the Assembly had little impact on federal policy, it served at least two other functions. It provided a means for mainstream women's organizations to demonstrate their patriotism in the heat of the Cold War; and using Cold War language, it enabled some of its member organizations to promote women's activism outside the home. Like the BPW and the AAUW, the Assembly called for "full utilization of the talents and capacities of women" and "maximum participation and use of womanpower in national security."[33] It continued to meet at least as late as 1958.[34]

Some policymakers also adopted this particular form of Cold War argument. As Alice Kessler-Harris has written, the Women's Bureau, a woman-run agency in the Department of Labor, used the "Cold War mentality" as "the wedge it needed to open some . . . [job training] programs" to women.[35] Under General Order 48, issued by the Secretary of Labor in 1950, the Women's Bureau assumed responsibility for "matters relating to the recruitment, training, and utilization of women for meeting defense and essential civilian labor requirements." Through the 1950s, it published bulletins and sponsored conferences, calling not only for training "for occupations which women ordinarily do not perform," but also using the nation's defense needs to argue explicitly in favor of equal pay, child-care facilities, opportunities for older women workers, and jobs for women in the sciences.[36] The Women's Bureau worked with the Office of Defense Mobilization and the Women's Advisory Committee on Defense Manpower, both of which also used Cold War language to encourage the use of "womanpower." In 1952, for example, on the urging of the Women's Advisory Committee, the Women's Bureau sponsored a two-day national conference endorsing equal pay. The keynote speaker Arthur S. Flemming, of the Office of Defense Mobilization, argued: "If we are going to utilize the womanpower of our Nation in an intelligent and efficient manner, and if our defense mobilization program is to rest on a solid foundation . . . then it is perfectly clear to me that this principle of equal pay for equal work and the principle of equal job opportunies for women are 'musts.'"[37] In large part because of the efforts of the Women's Bureau, the issue of equal pay remained on the national political agenda until Congress passed the federal Equal Pay Act in 1963.[38]

Historian Susan Hartmann has uncovered similar Cold War language in quasi-official policy organizations of the postwar era. In the 1950s, both the Commission on the Education of Women, sponsored by the American Council on Education, and the National Manpower Council, funded by the Ford Foundation, promoted education and employment for women by using Cold War arguments about the nation's security needs.[39] The organizations sponsored research and conferences, and offered policy recommendations to government, educational institutions, and employers. In 1957 the National Manpower Council published its influential report, *Womanpower.* It advocated expanded job training, placement services, and higher education for women, and it recommended that "employers hire, assign, train, and promote all individuals regardless of sex." The authors based their recommendations directly upon the premise that the "nation's strength and security depend upon its manpower resources."[40]

The arguments about national defense needs were not used only on behalf of women. Such arguments were used most effectively perhaps to urge federal funding for science and education,[41] but even here they had their impact on women. In 1957 the President's Committee on Scientists and Engineers recommended breaking down "employment barriers to women in science, engineering, and the technician fields."[42] In 1958 the National Defense Education Act authorized millions of federal dollars for grants, fellowships, and loans to enhance American education, money that funded women as well as men.[43]

In 1961, when President Kennedy issued Executive Order 10980 establishing a new Commission on the Status of Women, he included the now-familiar justification: "It is in the national interest to promote the economy, security, and national defense through the most efficient and effective utilization of the skills of all persons."[44] Activists from the BPW, AAUW, Women's Bureau,

Commission on the Education of Women, and National Manpower Council served in various capacities on the President's Commission. In this way, the activists who had developed the Cold War arguments for reform in the 1950s helped launch the women's movement of the 1960s.

Despite the conservative climate of the postwar era, a movement for sexual freedom also made headway during the 1950s. *Playboy* and other girlie magazines, pulp paperbacks, and the poetry of the Beats pushed the boundaries of acceptable sexual expression and won support in court decisions loosening the bonds of censorship.[45] The American Civil Liberties Union joined with publishers, authors, nudists, and others to demand more liberal approaches to sexuality. As part of this broad unorganized movement for greater sexual freedom, "homophiles," as early gay activists called themselves, came to the fore as champions of homosexual rights.

For the homophiles, the connections between the Cold War and homosexuality were immediate and direct. As John D'Emilio and others have shown, in the early 1950s conservative politicians and other avid red-baiters attacked not only communism but also "sex perversion," which they labeled a moral menace and national security risk. In response, a Senate subcommittee report and eventually an Executive Order called for federal officials to rout gay men and lesbians from government office. Meanwhile, in cities across the nation, local authorities harassed and arrested suspected homosexuals. In this overtly hostile climate, Harry Hay, a gay member of the Communist Party, joined with four other left-leaning men to organize the Mattachine Foundation, a Los Angeles–based group devoted to homosexual civil rights.[46]

Although Hay soon quit the Communist Party, his leftist past continued to haunt him. In 1953 a Los Angeles newspaper article revealed that the Mattachine's lawyer had appeared as an unfriendly witness before the House Committee on Un-American Activities. With this news, more conservative members of the growing organization spearheaded a political "cleansing" to rid it of the communist taint. In mid-1953, the anticommunists took over leadership of the Mattachine Society and steered the group away from its radical roots.[47]

By mid-decade the homophile movement had two major magazines: *ONE: The Homosexual Magazine,* first published in January 1953, and the *Mattachine Review,* the official journal of the Mattachine Society, first published in January 1955.[48] Both magazines grew out of the homophile movement in Los Angeles, and both, it seems, published articles written primarily by white, middle-class, gay men, with a smattering of articles by lesbians. The magazines, though, differed somewhat in their politics. From its beginnings, *ONE* took a more militant stance on issues of homophobia and also refused to engage in anticommunist purges. In the mid-1950s, some of the original leftist homophile leaders continued to write for *ONE*.[49] In contrast, the *Mattachine Review* adopted an unequivocally anticommunist stance. On several occasions the magazine reprinted the organization's anticommunist declaration: "The Mattachine Society is . . . unalterably opposed to Communists and Communist activity and will not tolerate the use of its name or organization by or for any Communist group or front."[50]

Despite their different politics, both *ONE* and *Mattachine Review* embraced certain kinds of Cold War language to promote their common cause. Like the women's movement, the *Mattachine Review* at least occasionally printed arguments about national security and wasted national resources. For example, James Barr Fugate, a WWII veteran discharged for homosexuality, asked: "Are we so rich in manpower that we can afford to discard thousands of able-bodied young men who are capable and willing to defend their country simply because they are sexual deviates?" He then pointed to the Soviet Union's numeric advantage with regard to military-eligible men.[51] But more often both magazines avoided such practical arguments about human resources and turned instead to more abstract Cold War argument.

In both homophile publications, the most common use of Cold War language applauded liberal individualism in a "free" society and contrasted it directly with "totalitarianist regimes such as that in communist Russia."[52] In this line of argument, the rights of homosexuals belonged within the broader battle for the kinds of individual rights allegedly protected in America and denied under communism. "In a democratic society," wrote one author, "when the minds of men are free, uniqueness and individualism can be signs of both personal and social health. But in totalitarian thought rigid conformity is the order of the day."[53] In part, such arguments echoed leftist social thought on authoritarian personalities, mass society, and the rise of the Nazis. Yet by the 1950s, the concept of totalitarianism—as developed by Hannah Arendt and others—conflated communism and fascism. As an author in *ONE* magazine put it, "the rights of the individual *must be* protected if we are not to become a police state such as . . . in Nazi Germany and . . . in Soviet Russia."[54] The Cold War rendition of totalitarianism thus enabled homophile activists to place the cause of gay rights on the side of "good" American ideals of individual freedom and to associate homophobic suppression with "bad" communism as well as fascism. An article in *ONE* stated the argument directly:

> The magnificence of the "American Ideal" lies in the discovery that the world, or any country in it, is large enough for differing peoples with different concepts of right

> and wrong to live together in tolerant harmony. The nation we most decry is that nation where the dead hand of conformity has done its worst, where all thought and action must fit what is officially acceptible [*sic*].[55]

In the homophile magazines, a few authors explicitly denounced the treatment of homosexuals in "totalitarian" societies. An article in *Mattachine Review,* for example, detailed the anti-gay policies of Stalinist Russia.[56] But more often, the magazines focused on the United States and the promise of democracy. In the *Mattachine Review,* one headline announced "Democracy Requires Dissenting Opinions."[57] For some homophile activists, homosexuals were a vanguard of spirited dissenters who resisted the looming threat of totalitarian moral authority. An article in *ONE* stated: "A dynamic, progressive society depends on those social groups, such as homosexuals, who are forever challenging the old ways, seeking a new outlook, and resisting the deadening hand of authoritarianism."[58] Gay men and lesbians, then, did not simply deserve democratic rights; by their very existence as dissenters, they strengthened American democracy against the perceived totalitarian threat. "We are the vanguard," an editorial in the *Mattachine Review* proclaimed, "of a movement that will someday widen and strengthen Democracy in the world."[59]

Other defenders of sexual freedom adopted similar language in the postwar era. Postwar artists and authors charged with obscenity, and the lawyers and social critics who supported them, routinely defended the right to free erotic expression as a mainstay of democracy and condemned censorship and sexual restriction as totalitarian techniques. In an editorial, the short-lived magazine *Sex and Censorship,* published in 1958, argued that "every dictatorship . . . has had censorship in one form or another." It then warned against the looming threat of American censors who "are not as free in this free country to impose their thinking on others as they would like to be." The censors, associated with foreign dictatorships, threatened freedom, associated with the United States.[60]

In their 1959 anticensorship classic *Pornography and the Law,* Eberhard and Phyllis Kronhausen stated the argument more directly: "The more actual democracy a society allows, the more sexual freedom is granted to its members. The more authoritarian the political organization of a society . . . the less sexual freedom." Lest readers miss the Cold War reference, the Kronhausens followed immediately with "illustrative examples" of sexual restrictions in the "Stalinist-type dictatorship" of the Soviet Union.[61] The homophile magazines belong within this more general postwar movement for freer sexual expression, a movement shaped by Cold War concepts of freedom and suppression.

Placed in a broader context still, the homophile activists stood with the many liberals who associated conformity and mass society with communism and asked Americans to live up to the ideals of a "free" society. Various commentators used the mythic battle between freedom and totalitarianism to legitimate abstract expressionist art, to critique conformism in suburbia, or to champion libertarian rugged individualism.[62]

In the climate of the postwar era, the Cold War arguments on sexual freedom had less impact on policymakers than did the more practical arguments on "womanpower." No federal agencies took up the banner of sexual freedom in the same way that they promoted women's contribution to the nation's defense. Still, the Cold War arguments on sexual freedom had their influence, especially in the courts. As the courts chipped away at censorship of sexually explicit books, magazines, and films, the lawyers, judges, and justices sometimes turned to Cold War language about freedom and suppression. On the U.S. Supreme Court, Justice William O. Douglas stood out as the most articulate spokesperson for the Cold War arguments. As early as 1946, in *Hannegan v. Esquire,* Douglas, delivering the opinion of the court, claimed: "Under our system of government there is an accommodation for the widest varieties of tastes and ideas . . . a requirement that literature or art conform to some norm prescribed by an official smacks of an ideology foreign to our system."[63] But he made his strongest case in 1961, in his dissenting argument in *Times Film v. City of Chicago:*

> Regimes of censorship are common in the world today. Every dictator has one; every Communist regime finds it indispensable. . . . Whether—as here—city officials or—as in Russia—a political party lays claim to the power of governmental censorship, whether the pressures are for a conformist moral code or for a conformist political ideology, no such regime is permitted by the First Amendment.[64]

In this view, homegrown American censorship of sexual expression was akin to political repression under communism, and both endangered freedom.

The movements for women's rights and sexual freedom were small though not inconsequential in the postwar era, faint counterpoints perhaps to the more conservative clamor. Most likely, these movements would have existed in different form without a Cold War culture; in fact, they did exist before the Cold War accelerated.[65] Even at the height of the Cold War, they did not rely solely on Cold War language to promote their goals.

The Cold War language of reform nonetheless illustrates a few of the ways that reformers adapted the Cold War to liberal causes. To dismiss such language as background "noise" or as desperate rhetorical flourishes adopted solely out of weakness is to dismiss the significant ways in which a complex Cold War culture shaped liberal reform as well as conservative politics. The Cold War arguments on the need for national resources and on freedom and suppression formed constitutive parts of postwar liberal ideological perspectives. They had at least some impact on reform, policy, and law in the postwar years, and they nurtured the seeds of change that sprouted in the decade that followed.

To study this Cold War language is not simply an exercise in reading "against the grain," in teasing out minority oppositional readings of an alleged Cold War consensus. Instead it offers lessons for future studies of gender, sexuality, and reform in the postwar era. First, the Cold War language of reform shows that the meanings of the Cold War were contested. Cold Warriors of various stripes might have agreed on their hatred of communism at home and abroad, but they did not necessarily agree on its broader meanings for American society. Anticommunist ideals could support traditional concepts of sex and gender, and they could also subvert them. What is sometimes presented as a one-sided cultural clampdown involved a multifaceted contest with coded debates we have only begun to decipher.

Second, the Cold War language of reform illuminates what Stuart Hall has called "the internal contradictions between those different ideologies which constitute the dominant terrain."[66] Reformers could and did exploit those contradictions; they used one ideal in their attempt to undermine another.[67] In the examples presented here, reformers reworked Cold War ideology to undermine ideals of domesticity and sexual containment. Using Cold War language, they promoted women's public participation, supported the Equal Rights Amendment, and argued for homosexual rights and more generally for sexual freedom. They reinscribed the ideology of the Cold War as they attempted to carve out respectable oppositional niches on gender and sexuality.

Other postwar reformers and social critics, no doubt, also used Cold War ideals to push for various forms of social change.[68] The Cold War destroyed the American left, but ironically it also provided vocabulary that helped some social movements survive and even grow in a conservative era. Instead of contrasting the conservative 1950s with the activist 1960s, recent historical writings have pointed to continuities, to roots of 1960s activism in 1950s movements.[69] In the women's movement and the movement for sexual freedom, and perhaps in other movements as well, those 1950s roots were grounded in a liberal variant of the Cold War culture.

Notes

1. Elaine Tyler May, *Homeward Bound: American Families in the Cold War Era* (New York: Basic Books, 1988); Elaine Tyler May, "Commentary: Ideology and Foreign Policy: Culture and Gender in Diplomatic History," *Diplomatic History* (Winter 1994);

Michael Paul Rogin, *Ronald Reagan, the Movie, and Other Episodes in Political Demonology* (Berkeley: University of California Press, 1987); John D'Emilio, "The Homosexual Menace: The Politics of Sexuality in Cold War America," in Kathy Peiss and Christina Simmons, eds., *Passion and Power: Sexuality in History* (Philadelphia: Temple University Press, 1989); Emily S. Rosenberg, "'Foreign Affairs' after World War II: Connecting Sexual and International Politics," *Diplomatic History* (Winter 1994); Geoffrey S. Smith, "National Security and Personal Isolation: Sex, Gender, and Disease in the Cold-War United States," *International History Review* (May 1992); Geoffrey S. Smith, "Commentary: Security, Gender, and the Historical Process," *Diplomatic History* (Winter 1994); Robert J. Corber, *In the Name of National Security: Hitchcock, Homophobia, and the Political Construction of Gender in Postwar America* (Durham: Duke University Press, 1993). Concepts of gender difference also help explain American Cold War foreign policy. Analysts of foreign policy find traditional concepts of gender hierarchy embedded within and shaping the content of Cold War diplomacy. See, for example, Cynthia Enloe, *The Morning After: Sexual Politics at the End of the Cold War* (Berkeley: University of California Press, 1993); Frank Costigliola, "'Unceasing Pressure for Penetration': Gender, Pathology, and Emotion in George Kennan's Formation of the Cold War," *Journal of American History* (March 1997).

2. Susan M. Hartmann, "Women's Employment and the Domestic Ideal in the Early Cold War Years," in Joanne Meyerowitz, ed., *Not June Cleaver: Women and Gender in Postwar America, 1945–1960* (Philadelphia: Temple University Press, 1994), 86.
3. The magazine changed its name to *National Business Woman* in 1957.
4. Leila J. Rupp and Verta Taylor, *Survivial in the Doldrums: The American Women's Rights Movement, 1945 to the 1960s* (Columbus: Ohio State University Press, 1990); see also Cynthia Harrison, *On Account of Sex: The Politics of Women's Issues, 1945–1968* (Berkeley: University of California Press, 1988).
5. See Rupp and Taylor, *Survival in the Doldrums,* 136, 138–139; Susan Lynn, *Progressive Women in Conservative Times: Racial Justice, Peace, and Feminism, 1945 to the 1960s* (New Brunswick, N.J.: Rutgers University Press, 1993), 195, n. 3.
6. On WILPF and the CAW, see Harriet Hyman Alonso, "Mayhem and Moderation: Women Peace Activists during the McCarthy Era" in Meyerowitz, ed., *Not June Cleaver;* Amy Swerdlow, *Women Strike for Peace: Traditional Motherhood and Radical Politics in the 1960s* (Chicago: University of Chicago Press, 1993), 37–40. On the Communist Party, see Deborah A. Gerson, "'Is Family Devotion Now Subversive?': Familialism Against McCarthyism," in Meyerowitz, ed., *Not June Cleaver;* Kate Weigand, "The Red Menace, the Feminine Mystique, and the Ohio Un-American Activities Commission: Gender and Anti-Communism in Ohio, 1951–1954," *Journal of Women's History* (Winter 1992).
7. Rupp and Taylor, *Survival in the Doldrums,* 136–41.
8. Lynn, *Progressive Women in Conservative Times,* 108. On women retreating from the left, see also Daniel Horowitz, *Betty Friedan and the Making of The Feminine Mystique* (Amherst: University of Massachusetts Press, 1998).

9. Rupp and Taylor, *Survival in the Doldrums;* Lynn, *Progressive Women in Conservative Times;* Nancy F. Gabin, *Feminism in the Labor Movement: Women and the United Auto Workers, 1935–1975* (Ithaca: Cornell University Press, 1990); Dorothy Sue Cobble, "Recapturing Working-Class Feminism: Union Women in the Postwar Era," in Meyerowitz, ed., *Not June Cleaver.*
10. On early BPW, see Judith Hole and Ellen Levine, *Rebirth of Feminism* (New York: Quadrangle/New York Times Book Company, 1971), 79–81; Nancy Cott, *The Grounding of Modern Feminism* (New Haven: Yale University Press, 1987), 89–90.
11. On leadership of the ERA campaign, see Rupp and Taylor, *Survival in the Doldrums,* 73.
12. Frances Maule, "Hightide," *Independent Woman* (September 1950), 262; Rupp and Taylor, *Survival in the Doldrums,* 26.
13. Rupp and Taylor, *Survival in the Doldrums,* 138–39; "The DAR Supplies Leadership," *American Mercury* (August 1958), 35; "National Federation of Business and Professional Women," *American Mercury* (February 1959), 155. The *American Mercury* editorial of 1958 also accused the League of Women Voters, the Parent-Teachers Associations, the American Association of University Women, and the YWCA. All were "too frequently counted in the ranks of the political left on vital issues. . . . The leadership seems always to have a tendency to drift into left-wing hands."
14. Rupp and Taylor argue that most women's rights activists of this era "were anti-Communists to one degree or another." Rupp and Taylor, *Survival in the Doldrums,* 136.
15. See, for example, Lisa Sergio, "Neighbors in One World," *Independent Woman* (October 1948), 285–86, 308; "Get into Politics Up to Your Ears," *Independent Woman* (November 1949), 333–34, 348. For later anticommunist articles, see Julia Cole Fauber, "They Can't Brainwash Me," *Independent Woman* (September 1954), 337–38; "The Long Way to Freedom," *National Business Woman* (February 1957), 2–4, 18.
16. "Total Defense—Our Best Chance for Peace," *Independent Woman* (August 1950), 245; "To Strengthen the Free World," *Independent Woman* (January 1951), 24; see also "Leaders Gather for Annual Forum," *Independent Woman* (November 1950), 352; "Program for Civil Defense Takes Form," *Independent Woman* (November 1950), 353, 359; Marjorie F. Webster, "Battleground U.S.A.," *Independent Woman* (July 1953), 239.
17. "Declaration of Principle," *Independent Woman* (August 1953), inside front cover.
18. On the pamphlet, see *National Business Woman* (March 1957), inside front cover.
19. "Total Defense," 245, 246.
20. Ella V. Ross, "Education and Vocations," *Independent Woman* (May 1951), 129, 130.
21. "Women Should Be Drafted," *Independent Woman* (April 1951), 113; "Our Stand on Registration and Draft of Women," *Independent Woman* (October 1951), 295; Sarah T. Hughes, "Our National President Asks Every Club to Help in November Drive to Recruit Women for Armed Services," *Independent Woman* (November 1951),333; Marjorie L. Temple, "We Frame a Bill for Registration and Draft of Women," *Independent Woman* (March 1952), 74.

22. "Resolutions," *Independent Woman* (August 1954), 297.
23. *National Business Woman* (February 1958), 3; "Tentative Legislative Platform—1958," *National Business Woman* (March 1958), 10.
24. "National Legislation Platform, 1958–1959," *National Business Woman* (September 1958), 20.
25. On African American civil rights and Cold War thought, see Mary L. Dudziak, "Desegregation as a Cold War Imperative," *Stanford Law Review* (November 1988).
26. Geneva F. McQuatters, "Elusive Equal Rights," *Independent Woman* (November 1950), 352.
27. "Federation Highlights," *Independent Woman* (October 1956), 28.
28. Assembly of Women's Organizations for National Security, *Bulletin* 1:4 (June 1951), 4–5, Somerville-Howorth Collection 7:147 (Schlesinger Library, Radcliffe College).
29. Dorothy W. Weeks, "Woman Power Shortage in the Physical Sciences," *Journal of the American Association of University Women* (March 1955), 146; Winifred Helmes, "Woman Power and Higher Education," *Journal of the American Association of University Women* (May 1958), 204.
30. "Total Defense," 246.
31. Margaret F. Ackroyd, "The Case against the Equal Rights Amendment," *Journal of the American Association of University Women* (October 1953), 27.
32. Letter from Sarah T. Hughes, President, National Federation of Business and Professional Women's Clubs, to Lucy Howorth, American Association of University Women, September 19, 1950, Somerville-Howorth Collection 7:147; "To Rally Women for Defense," *Independent Woman* (October 1950), 325; "Program for Civil Defense Takes Form," *Independent Woman* (November 1950), 353; see also Rupp and Taylor, *Survival in the Doldrums,* 78–79.
33. "Final Report of the Chairman, Lucy Somerville Howorth, of the Assembly of Women's Organizations for National Security, July 1951–February 1952," Somerville-Howorth Collection 7:147; Dorothy G. Stackhouse, "Assembly of Women's Organizations for National Security," *General Federation Clubwoman* (January 1954), 10.
34. "Woman's Assembly Opens at 2012," *National Business Woman* (November 1958), 9.
35. Alice Kessler-Harris, *Out to Work: A History of Wage-Earning Women in the United States* (New York: Oxford University Press, 1982), 304.
36. "Employment of Women in an Emergency Period," *Women's Bureau Bulletin* 241 (Washington, D.C.: Government Printing Office, 1952), 12, 8. For examples of the national defense argument, see also "Report of the National Conference on Equal Pay, March 31 and April 1, 1952," *Women's Bureau Bulletin* 243 (Washington, D.C.: Government Printing Office, 1952); "Employed Mothers and Child Care," *Women's Bureau Bulletin* 246 (Washington, D.C.: Government Printing Office, 1953); "Employment Opportunities for Women in Professional Engineering," *Women's Bureau Bulletin* 254 (Washington, D.C.: Government Printing Office, 1954); "Training Mature Women for Employment," *Women's Bureau Bulletin* 256 (Washington, D.C.: Government Printing Office, 1955); "Careers for Women in the Physical Sciences,"

Women's Bureau Bulletin 270 (Washington, D.C.: Government Printing Office, 1959).

37. "Report of the National Conference on Equal Pay," 13. Under General Order 48, issued in 1950, the Secretary of Labor created the Women's Advisory Committee on Defense Manpower.
38. For a good history of equal pay legislation, with an emphasis on the Women's Bureau in the postwar era, see Alice Kessler-Harris, *A Woman's Wage: Historical Meanings and Social Consequences* (Lexington: University Press of Kentucky, 1990), chap. 4.
39. Hartmann, "Women's Employment and the Domestic Ideal."
40. National Manpower Council, *Womanpower* (New York: Columbia University Press, 1957), 5, 7.
41. On funding for science, see Daniel Lee Kleinman and Mark Solovey, "Hot Science/Cold War: The National Science Foundation after World War II," *Radical History Review* (Fall 1995); on funding for education, see Barbara Barksdale Clowse, *Brainpower for the Cold War: The Sputnik Crisis and National Defense Education Act of 1958* (Westport, Conn.: Greenwood Press, 1981).
42. "Careers for Women in the Physical Sciences," 1.
43. For a brief reference to NDEA's impact on women, see Blanche Linden-Ward and Carol Hurd Green, *Changing the Future: American Women in the 1960s* (New York: Twayne, 1993), 68.
44. Margaret Mead and Frances Bagley Kaplan, eds., *American Women: The Report of the President's Commission on the Status of Women and Other Publications of the Commission* (New York: Charles Scribner's Sons, 1965), 207.
45. On the postwar shift in judicial approaches to censorship, see Edward De Grazia, *Censorship Landmarks* (New York: R. R. Bowker, 1969), Introduction.
46. John D'Emilio, *Sexual Politics, Sexual Communities: The Making of a Homosexual Minority in the United States, 1940–1970* (Chicago: University of Chicago Press, 1983), chaps. 3, 4. See also Jonathan Katz, *Gay American History: Lesbians and Gay Men in the U.S.A.* (New York: Avon, 1976), 139–87, 611–25.
47. D'Emilio, *Sexual Politics, Sexual Communities,* chap. 5; Katz, *Gay American History,* 625–32.
48. A third magazine, *The Ladder,* published by a lesbian rights group, Daughters of Bilitis, first appeared in October 1956. I have not used that magazine because it rarely used Cold War language in the 1950s.
49. D'Emilio, *Sexual Politics, Sexual Communities,* 80.
50. *Mattachine Review* (January 1956), 13; see also *Mattachine Review* (January–February 1955), back cover; *Mattachine Review* (May–June 1955), 2.
51. James (Barr) Fugate, "Under Honorable Conditions," *Mattachine Review* (May–June 1955), 42.
52. *Mattachine Review* (July 1957), 2.
53. Carl B. Harding, "Deep Are the Roots," *Mattachine Review* (March–April 1955), 7.
54. Marlin Prentiss, "Are Homosexuals Security Risks?" *ONE* (December 1955), 6.

55. Lyn Pedersen, "An Open Letter: Do Constitutional Guarantees Cover Homosexuals?" *ONE* (January 1956), 11. Pedersen was one of the pseudonyms of James Kepner; D'Emilio, *Sexual Politics, Sexual Communities,* 111.
56. Philip Jason, "Progress to Barbarism," *Mattachine Review* (August 1957), 18–21.
57. "Democracy Requires Dissenting Opinions," *Mattachine Review* (March–April 1955), 25. This article reprinted a *Time* magazine column written by retired judge Learned Hand. The opening sentence read: "Why is it that totalitarianisms arouse our deepest hostility?"
58. "Six Roundtables," *ONE* (March 1958), 9.
59. "Editorial," *Mattachine Review* (March 1957), 2. An earlier version of this argument appeared in 1951. In *The Homosexual in America,* the pseudonymous Donald Webster Cory portrayed homosexuality as a "progressive" force in the battle against totalitarianism. Cory discussed the persecution of homosexuals in the "totalitarian" states of Nazi Germany and Soviet Russia. He concluded that homosexuality "will broaden the base for freedom of thought and communication, will be a banner-bearer in the struggle for liberalization of our sexual conventions, and will be a pillar of strength in the defense of our threatened democracy." Donald Webster Cory, *The Homosexual in America* (New York: Greenberg, 1951), 235.
60. Editorial, *Candida* 1:3 (1958), 20; see also Dale Hart, "Private and Public Censorship," *Candida* 1:3 (1958), 50. After its first two issues, *Sex and Censorship* was renamed *Candida.* It then ceased publication.
61. Eberhard and Phyllis Kronhausen, *Pornography and the Law: The Psychology of Erotic Realism and Pornography* (New York: Ballantine, 1959), 283–84. The Kronhausens praised the liberal sexual legislation of the U.S.S.R. under Lenin and contrasted it with the repressive dictatorship under Stalin.
62. On the development and use of this Cold War dualist (free society vs. totalitarian society) vision among liberals, see Richard Pells, *The Liberal Mind in a Conservative Age: American Intellectuals in the 1940s and 1950s* (New York: Harper and Row, 1985); Richard Gid Powers, *Not Without Honor: The History of American Anticommunists* (New York: Free Press, 1995). On art, see Erika Doss, *Benton, Pollock, and the Politics of Modernism: From Regionalism to Abstract Expressionism* (Chicago: University of Chicago Press, 1991), esp. chap. 6. (Other anticommunists explicitly denounced abstract art as subversive. See Jane DeHart Mathews, "Art and Politics in Cold War America," *American Historical Review* [October 1976].)
63. *Hannegan v. Esquire* 327 US 157–58 (1946).
64. *Times Film Corp. v. City of Chicago et al.* 365 US 79–80 (1961).
65. I refer here to the women's movement and the broad movement for sexual freedom. The American homophile movement did not exist in organized form before the Cold War; it emerged during the Cold War, in part in response to the Cold War assault on gay men and lesbians in government.
66. Stuart Hall, "Culture, Media, and the 'Ideological Effect,'" in James Curran, Michael Gurevitch, and Janet Woollacott, eds., *Mass Communication and Society* (Beverly Hills: Sage, 1979), 346.

67. Just as Cold War language could undermine traditional sex and gender ideals, so, conversely, could traditional concepts of family and gender undermine Cold War ideology; see Gerson,"Is Family Devotion Now Subversive?"
68. See, for example, Dudziak, "Desegregation as a Cold War Imperative."
69. See W. T. Lhamon Jr., *Deliberate Speed: The Origins of a Cultural Style in the American 1950s* (Washington, D.C.: Smithsonian Institution Press, 1990); Maurice Isserman, *If I Had a Hammer . . . The Death of the Old Left and the Birth of the New Left* (New York: Basic Books, 1987); Todd Gitlin, *The Sixties: Years of Hope, Days of Rage* (New York: Bantam, 1987); Margaret Rose, "Gender and Civic Activism in Mexican American Barrios in California: The Community Service Organization, 1947–1962," in Meyerowitz, ed., *Not June Cleaver;* Barbara Ehrenreich, *The Hearts of Men: American Dreams and the Flight from Commitment* (Garden City, N.Y.: Anchor Press/ Doubleday, 1983); Wini Breines, *Young, White, and Miserable: Growing Up Female in the Fifties* (Boston: Beacon Press, 1992); D'Emilio, *Sexual Politics, Sexual Communities;* Lynn, *Progressive Women in Conservative Times;* Rupp and Taylor, *Survival in the Doldrums.*

CHAPTER 6

Containment at Home

Gender, Sexuality, and National Identity in Cold War America

Jane Sherron De Hart

The baby boom, Barbie dolls, and the Beats seem to be as unrelated to the Truman Doctrine, NATO, or the Korean War as they are remote from contemporary debates about abortion, same-sex marriages, the proper role of Hillary Rodham Clinton as First Lady, or the military's "don't ask, don't tell" policy. Yet those icons of the 1950s can provide not only insight into family life, gender roles, and sexual behavior during the early years of the Cold War, but clues as to how these intimate matters were linked to national security in an era of superpower rivalry. Did the Cold War affect thinking about gender and sexuality in the United States? Satisfying answers remain tantalizingly elusive. The end of superpower hostilities invites us to think afresh about the question.

How we think about the linkage between international politics and the politics of gender and sexuality owes much to those students of American culture who first identified the possible connections. Elaine Tyler May, in her pathbreaking study of white middle-class family life, has been especially influential. In *Homeward Bound,* May linked the exaggerated domesticity and highly politicized homophobia that characterized the "long fifties" (from 1945 to around 1965) to anticommunist imperatives. She readily acknowledged the extraordinary stresses placed upon the American family by the Great Depression and World War II. A

For extensive versions of this chapter, see Jane Sherron De Hart, *Litigating Equality* and *Defining America* (University of Chicago Press, forthcoming).

depressed economy had meant aborted careers, delayed marriages, and deferred children. War had further disrupted traditional gender arrangements and sexual norms as women moved into the job market and the military in unprecedented numbers, experiencing new vocational and sexual opportunities that heralded new roles and identities. Postwar Americans, finding additional threats to traditional family life in rising rates of out-of-wedlock pregnancy, sexual promiscuity, juvenile delinquency, and the ultimate threat of nuclear war, had understandably sought "normalcy" in marital sex, pronatalism, and suburban domesticity. But the rigid heterosexuality and strict adherence to traditional gender roles promoted during the Cold War years, May insisted, involved more than a return to "normalcy." They constituted a domestic version of containment. Just as anticommunism required the containment of Sino-Soviet expansion abroad, so, she argued, gender revolution and deviant expression of sexual desire had to be effectively contained at home. Promotion of family values, policymakers believed, would assure the stable family life necessary for personal and national security as well as supremacy over the Soviets. No mere exercise in nostalgia, domestic containment was part of a new Cold War consensus about the meaning of America.[1]

May's use of domestic containment as a metaphor for Cold War gender and sexual politics is not without problems, for reasons we will explore. Yet the repression that she identified was real and has been underscored by scholarship on the gay and lesbian experience during the long fifties. The gay community, which had emerged with new cohesion and visibility in the wake of World War II, found itself a prime target for anticommunist crusaders. Homosexuals, it was believed, were especially vulnerable to blackmail by Soviet agents eager to recruit intelligence sources. As a result, the early 1950s witnessed widespread purging of homosexuals from the State Department, the military, and other federal agencies. More than just a greater risk of blackmail was involved. Homosexuals were seen as deficient in character, moral integrity, and real masculinity. Unfit as Cold Warriors, they were undesirable citizens. The stigmatization, as historian John D'Emilio points out in his study of Cold War sexual politics, was carried still further by conservative politicians and "right-wing ideologues" who linked homosexuality to communism, reconceptualizing homosexuality as a contagious disease spread by communists to weaken the nation from within. A single homosexual, officials maintained, could easily contaminate an entire government office. The oppression of homosexuals at the federal, state, and municipal levels became yet another act of containment in the fight against communism. Gay federal and state employees, military personnel, university professors, and workers for government subcontractors were scrutinized, investigated, and sometimes fired for suspected homosexual activity.[2]

Purging homosexuals and lesbians from public institutions was only the first step, however. Despite Alfred C. Kinsey's findings in his 1948 report that 37 percent of American males had engaged in post-adolescent same-sex sexual activity at least once, homosexuality was further pathologized, and gays and lesbians were labeled "sex deviates" and "perverts" by government officials, psychologists, and journalists alike. As marriage—and female sexuality—became increasingly eroticized during the fifties, homosexuality was further demonized in order to create the boundaries for acceptable sexuality. If "normal" sexuality was contained within the home and heterosexual marriage, then homosexuality, which could only occur outside these boundaries, threatened the social order, as did any extramarital sexuality. Raids on gay bars, the conflation of lesbians and prostitutes, media exposure of sexual transgressors, and the demonization of homosexual men as child molesters functioned to shore up social controls on sexual behavior. The state's increasing intervention into the realm of sexuality represented an effort both to restrict women and female sexuality within the confines of home and marriage and to protect society from "sex deviates" by removing homosexuals and other sexual transgressors from the public sphere.[3]

In this endeavor, state action was reinforced by unofficial boundary keepers. Literary and cultural critics sought to contain the emergence of a distinctly gay male literary tradition by assimilating the work of Tennessee Williams, Gore Vidal, and James Baldwin into the dominant paradigms of American fiction. Hollywood also cooperated. Consistent with its self-imposed censorship code, the film industry kept homosexuals in the closet, prohibing explicit treatment of homosexuality in films. To the extent that a gay male character was included, he was made virtually indistinguishable from straight male characters, thereby demonstrating that homosexuals, like communists, were especially dangerous because they could so easily escape detection. Even film noir, which critiqued the dominant order in other respects, furthered containment in its presentation of the femme fatal and gay male. While the gay male almost invariably merged as the psychopathic killer, the femme fatal, by refusing to restrict her sexuality to the domestic sphere, allowed her transgressive behavior to become as much a target of investigation by the hard-boiled detective hero as that of the victim, notes cultural critic Richard Corber. Best known for "selling" of domesticity along with household products was the new television industry. Shows such as *The Adventures of Ozzie and Harriet, Father Knows Best, The Donna Reed Show,* and *Leave It to Beaver* purveyed containment along with homogenized, romanticized views of contented Moms who never know as much as Dad.[4]

But before we conclude that "domestic containment" is the correct descriptor for this system of controls, there are many questions to be explored. First, what precisely is meant by the term? Was this thicket of restraints a new Cold War offshoot or was it old vegetation uprooted by war and replanted in the fertile soil of a thriving postwar mass-consumption economy? To put the question differently, would "domestic containment" have existed without the Cold War, perhaps by another name? Did the gender and sexual norms that so unrelentingly bombarded the middle-class whites whom May studied also dictate behavior of Americans who were neither white nor middle-class? Can the resurgent domesticity of the postwar era be reconciled with both a growing female presence in the labor market during the long fifties and newly documented activism in the public sphere?

What about the women's and gay liberation movements of the 1960s? Are they related to the thaw in the Cold War and the breakdown of containment policy in Vietnam? Did the liberation movements of the 1960s succeed in replacing the coercive gender and sexual norms? If so, how do we think about the remaining decades of the Cold War? Are we talking about a plurality of discourses in the 1970s and 1980s? Did liberal defection with respect to containment at home and abroad during the Vietnam years, and loss of state support for strict policing of gender and sexual boundaries, intensify conservative determination to maintain those boundaries? Finally, given the demise of communism in Eastern Europe and the breakup of the Soviet Union, how do we explain the renewed insistence on boundaries in the gender and sexual agenda of conservative politicians and their constituents in the 1980s and the 1990s?

However we answer these questions, the binary opposites of containment and liberation, even if used sequentially, seem much too simple to capture the complex realities of gender and sexuality over nearly half a century. Even if we lack definitive answers, the questions themselves suggest a need to rethink our periodization and conceptualization of gender and sexual politics and the Cold War.

The purpose of this chapter is to stimulate such efforts. My intent is not to provide a detailed overview of changing relationships of sexuality and gender since the end of World War II. Nor do I compare the gender and sexual politics of the Cold War years with that of earlier periods in American history, although such comparisons would be informative. Rather my objectives here are more limited. The first section probes whether domestic containment is indeed an appropriate metaphor for the sex/gender system prevailing in the early years of the Cold War, assessing key findings in recent scholarship. The second section, after briefly tracing the emergence of liberation movements of the 1960s, focuses on the conservative response to challenges posed by the erosion of containment at home and abroad. The third and final section examines why the gender and sexual politics of the early Cold War years has outlasted the threat of international communism to which it was so intimately linked. The answer, I submit, requires viewing domestic containment in the larger context of constructions of American national identity.

Examining domestic containment in relation to national identity has multiple advantages. It allows us to see that what we have heretofore regarded as a unique Cold War phenomenon was in fact part of the larger, ongoing process of defining America. The Cold War may have given this particular construction of national identity its distinctive coloration. But once the specter of international communism faded and old fissures within American society widened, fundamental questions about national identity and purpose once again pushed to the surface. That the Cold War has given way to the culture wars suggests how problematized the very meaning of America has once more become.

So much of this current cultural conflict revolves around issues such as abortion, gay rights, and unwed welfare mothers, themselves a measure of just how embedded both gender and sexuality remain in constructions of American identity. Finally, the intensity with which those conflicts are waged is a measure of how deeply the bipolar categorization and moral absolutism of the Cold War continues to mark our civic discourse.

Our first task is to determine whether domestic containment is actually an accurate descriptor of gender and sexual ideology and behavior. The term, as May intended, situates intimate matters within Cold War discourse, making containment of incipient gender and sexual liberation at home analogous to containment of international communism and Sino-Soviet expansion abroad. The parallels, however, are not exact. Containment with respect to foreign policy as originally conceived was intended to block expansion of a foreign power with which the United States would coexist. On the other hand, the stigmatization and repression associated with postwar gender and sexual policing, particularly the latter, hardly implied peaceful coexistence. What it did imply was rollback—something that for all the harsh rhetoric of Cold Warriors such as John Foster Dulles was never attempted abroad, with the critical exception of General Douglas MacArthur's invasion of North Korea. Despite these discrepencies, a likeness between containment at home and abroad may be conceded to reside metaphorically in the idea of boundary maintenance. But there is more to consider.[5]

As May herself acknowledged, the economic and military exigencies of World War II were deeply implicated in the blurring of conventional gender roles and the maximizing of opportunties to act on same-sex as well as opposite-sex desires. With tradition and virtue withering under fire, might efforts to reconstruct old boundaries and identities have occurred with demobilization even if there had been no Cold War?

The question, while counterfactual, should not be dismissed. Modern wars provide battlegrounds where sexual and gender systems are as vulnerable to explosion as are the men in uniform. Historians of the interwar years have reminded us that in the wake of World War I, efforts in Britain and France to restabilize gender hierarchies and identities, repopulate the homefront, and repress sexually transgressive behavior took a variety of forms. Even allowing for national variations, these measures and the anxieties which fed them have a familiar ring: removal of women from wartime jobs, pronatalist legislation, attacks on feminism, opposition to the "new woman"—"this being without breasts, without hips"—concern about the single woman "destined" to remain unmarried because of war-inflicted mortality and the resulting uneven sex ratio, the need of veterans to reassert virility through paternity, new emphasis on conjugal sex as a vehicle for sexual pleasure, the emergence of a *Kinder, Küche, Kirche* ideology stressing femininity and motherhood. So the list goes on. It is clear that in the post–World War I period, too, discourse about gender and sexuality, whether articulated in public policy or cultural narratives and images, was in part about reinstating boundaries.[6]

Boundary reconstruction may well be a common postwar theme. But more may be involved. The post–World War II American variation was most definitely a Cold War phenomenon in its specifics. The impetus may have derived from wartime disruption. It was powered, as May suggests, by the specter of international communism and nuclear annihilation. Fear of communism permeated American life in the long fifties. By no means concocted entirely of phantasms, it was rooted in legitimate concerns about Stalinism and its expansion as well as breaches in national security that allowed the Soviets' access to classified information. Yet the dangers posed by communism were *to* the United States, rather than *in* the United States—a point made by the literary critic Philip Rahv in the 1950s. That distinction escaped those Americans who threw themselves into the anticommunists crusades of the McCarthy era with scant concern for the costs involved.[7]

Conditioned by a long history of antiradicalism and an earlier Red Scare in which patriachy and patriotism were ideologically conflated, they had become accustomed in the 1920s to associating communism with gender upheaval and sexual transgressions. In the face of new threats at home, such linkages once again seemed credible. Supported structurally by an expanding mass-consumption economy, the new gender and sexual politics of the long fifties provided Cold War liberals and conservatives alike a way to maintain the fiction of a stablitity in intimate matters that so eluded them in national and international affairs. If in both the domestic and the foreign arena, militancy overshadowed peaceful coexistence, distorting containment in its original meaning, that too was part of Cold War dynamics.[8]

The term "domestic containment," therefore, can legitimately be used despite its problematic character. For purposes of this chapter it serves as a powerful and generic metaphor for strenuous, systematic efforts to maintain traditional gender and sexual boundaries, ideologically as well as behaviorally, through cultural imperatives and social policy.

Having concluded that domestic containment is an appropriate metaphor, we need to determine to whom these gender and sexual norms applied. To do so requires extending our scrutiny beyond the middle-class white families that were the subject of May's study to women and men who were working class, black, or Hispanic. Since the publication of *Homeward Bound* in 1988, these groups have been the focus of new scholarship which, while hardly definitive, does enlarge our understanding of the social and ideological parameters within which individuals made private and public choices during the long fifties. More important for our purposes, this more recent scholarship, much of it about women, illuminates the extent to which choices were justified in ways that reaffirmed or contested the prevailing gender and sexual norms associated with domestic containment.

Some of this newer work provides additional documentation of behavioral conformity to containment ideology. The women whom Brett Harvey interviewed for her oral history of the 1950s, for example, did what they were supposed to do: they dropped out of college or chose not to pursue graduate degrees in order to get married; they often married in order to have sex, fearing the consequences of premarital relationships, especially pregnancy; they often chose partners who they anticipated would be good providers; and they had children because they felt they were supposed to, staying home with them despite the boredom that, for many, childrearing and housekeeping produced.[9]

Yet if many women acted within the confines of containment, other women behaved in ways that at first glance seem to be at odds with the long-dominant model of domesticity and political quiescence. Historians have forever noted the increasing numbers of women, many of them married and mothers, who moved into white-collar and clerical jobs in the postwar years. New studies confirm the difficulties such women had reconciling the dictates of domesticity with the reality of increasing employment. Avoiding public justification of their employment in terms of intellectual stimulation, personal satisfaction, or financial gain, they used the familial rhetoric of obligation and self-sacrifice, although many middle-class women privately confessed that working was something they did to "stay sane." For example, married nurses, responding to a manpower shortage in the 1950s, framed their return to the work force as a public service and personal sacrifice. Their families, they emphasized, suffered no neglect as a result of their employment.[10]

Even organizations like the National Manpower Council and the National Council on the Education of Women that worked to change public opinion about married women's workforce participation insisted that women's first priority was home and family. Employment was justified within the framework of the female life cycle: women would work until they got married and after their children were all in school. Congress, in debating the Internal Revenue Act of 1954, also refused to acknowledge formally women's right to work outside the home. Advocates of child-care deductions for working women argued in terms of expediency (i.e., teachers and nurses were vital to the nation welfare; poor women who worked must do so to avoid the greater evil of welfare dependency; and child care reduced juvenile delinquency).[11]

Women who left the home, not as workforce participants but as political activists, also maneuvered within the framework of containment even as they challenged the basic assumptions of defense and racial policy. Advocates for international peace and civil rights took very real risks as they protested against the nuclear arms race and the Korean War or agitated against segregation of blacks and for civil liberties for victims of McCarthyism. But they clothed their actions with gendered rhetoric and symbols. The American Women for Peace, for example, invoked essentialism, arguing that the process of giving birth meant that women had a "natural responsibility" to preserve life. Members of the Civil Defense Project and the Women's Strike for Peace brought their children with them to demonstrations: babies provided tangible evidence of motherhood and a deterrent to arrest.[12]

Racial segregation also brought both black and white women into the public arena where dissent was shielded in gender conformity. When racial integration generated threats to shut down public schools in the South, the white women of Atlanta who lobbied for open schools couched their demands in maternalist language that emphasized their responsibility as mothers for the education of "all children," not just their own. Women of the Young Women's Christian Association (YWCA) and the American Friends Service Committee challenged segregation as sisters in Christ, invoking a female ethic that focused on nurturing personal relationships and networks of support across racial lines.[13]

Even women at the margins of society recognized the power of containment discourse. Wives of Communist Party members arrested under the Smith Act clothed their actions on behalf of their husbands' release in familial language, depicting McCarthyism as an attack on the family. Women of color whose racial and class status had previously exempted them from dominant constructions of motherhood also politicized the very constructions from which they had been excluded. Mamie Bradley was a case in point. The mother of Emmett Till, the black teenager who had been murdered by two white men while vacationing in Mississippi for allegedly whistling at the wife of one of his assailants, sought justice for her son's murder. To the extent that the assailants' conviction required Bradley to represent herself as a respectable middle-class woman and good mother, the grieving Bradley did her best. Mexican-American women, working through barrio-based Community Service Organizations, also couched their agitation for neighborhood improvement, health care, and education as an extension of their roles as mothers, although their emphasis on domesticity and family may have been as much the product of ethnic tradition as Cold War dictates.[14]

In an era when dissent was routinely equated with communism, it is hardly surprising that so many women whose actions challenged containment norms used essentialist, maternalist, and familial rhetoric to justify their behavior. The practice was a venerable one, having been resorted to by women reformers seeking to legitimate public activism throughout the nineteenth and early twentieth centuries. Some of these modern practitioners of maternalism were prepared to admit privately that motherhood and family were not the only basis for their political involvement: egalitarian considerations also played a part.[15]

Others, who were less tightly linked to the dominant culture, were more candid, deferring only minimally to containment ideology in crafting a rationale for activism. Black women working for integrated schools defended their actions in terms of racial justice as well as maternal obligation. Trade union women, while insisting that being a good mother also meant providing financially for one's children, pushed for collective bargaining, pay equity, maternity leave, and child care as "citizens and workers." Members of leftist organizations such as the Congress of American Women were even less equivocal, omitting references to maternal or familial responsibilities completely.[16]

While domestic containment—like any other hegemonic discourse—generated contestation (both covert and overt), the extent to which Americans drew from this rhetorical wardrobe in order to cloak nonconformist behavior was a measure of its power. Another more explicit measure was the greater scrutiny and sanctions applied to maintain official gender and sexual boundaries. Some of those sanctions appeared as negative linkages: characterization of the "enemies of American society as "lecherous," of student radicals as sexually "depraved." Other sanctions took the form of new conceptualizations: the redefinition of lesbianism as a "deviant" expression of female sexuality, of homosexuality as "mental illness," and of black illegitimacy as "cultural pathology." Still other proscriptions appeared as prescription: intensified admonitions to parents to counter incipient tendencies toward homosexuality in "sissy" boys and "unhappy" tomboyish girls; urgent warnings to women to prevent sex crimes by staying home and keeping their children off the streets; heightened pressure on coaches and colleges to disavow signs of mannishness in female athletes and on female sports figures to demonstrate femininity and heterosexuality; pressure on working-class "butch" lesbians by other lesbians to tone down their masculinized personae so as not to invite charges of gender deviance.[17]

Coercion took a more official form with increased municipal crackdowns on the sale of "indecent" magazines; censorship of lyrics in rock and roll music; denial of child custody to lesbian mothers; heightened police surveillance of gay and lesbian bars, parks, and other public spaces where homosexuals congregated; increasingly public prosecution of female abortionists as violators of womanhood—women who destroyed other women's fetuses and motherhood for profit. Regardless of how they were packaged, all these actions were designed to reconstruct traditional gender roles and expand social control over sexuality by psychiatry and the state.[18]

Race was a mediating factor in determining to whom the controls applied. Nowhere was this more evident than the music world. For example, suggestive lyrics went uncensored in Joe Turner's rhythm and blues version of "Shake, Rattle, and Roll" that was played on black radio stations in 1954. However, as large numbers of white teenagers begin listening to black rhythm and blues and as white performers like Elvis Presley began to adopt rhythm and blues techniques, censorship became the norm. When Joe Haley recorded his version for a white audience that same year, the sexuality was muted. As Haley explained, "we steer clear of anything suggestive." Policing sexual boundaries was also a way of policing racial boundaries. When rhythm and blues shows were presented before mixed audiences on television in Los Angeles in the early 1950s, local authorities were on hand immediately. As the shows' producer recalled, "The cops would come and hassle the kids standing in line to get into the television shows. If it were all Asians and Hispanic and black [kids], they wouldn't care, but there were whites there, and they're mixing with the blacks. . . ." Although laws preventing racial intermarriage were beginning to be relaxed in the western states, barriers to a mestizo America had to be maintained.[19]

The results were never so thoroughgoing as intended by the architects of domestic containment—those policymakers, public officials, and professionals who fanned the home fires in the national-security state. Contestation took many forms: articles in mass circulation magazines for women who celebrated not only domesticity but also individual achievement in the public sphere; the female subculture of "swinging singles" that Helen Gurley Brown would publicize in *Sex and the Single Girl;* the emergence of Hugh Hefner's "playboy philosophy" with its promise of sexual pleasure and escape from the responsibilities of conventional marriage and parenthood; the persistence of a thriving working-class lesbian bar culture in the nation's port cities where many lesbians remained after their discharge from the military; a small "homophile" movement in California, one of the few states where homosexuals could legally congregate in bars and other public spaces; white hipsters who sought in bebop music a new personal identity; the highly visible beat cultures of North Beach and Greenwich Village with their heady mix of sex, drugs, and adventure. And, by 1960, there was the pill with its promise of risk-free sexual gratification. That these and other precursors of revolution were gathering in the fifties and early sixties suggest a transition in the making. But harbingers of change should not obscure the reality of gender and sexual containment—or the tenacity of its appeal, especially to social conservatives for whom the gender, sexual, and racial norms and behavior associated with the early years of containment were not a matter of social policy but of personal identity.[20]

In a society in which the media created instant awareness of social change, feminism and gay liberation burst upon the public consciousness with all the éclat of a fireworks display on the Fourth of July. The more radical elements of the women's movement, with their talk of test-tube conception, the slavery of marriage, and the downfall of capitalism, might have been dismissed out of hand. But it was hard to ignore 50,000 women parading down New York's Fifth Avenue, *Ms.* magazine on newsstands, feminist books on the best-seller lists, coeds in the nation's military academies, women in the ministry and rabbinate, or even women in hard hats on construction jobs. It was still harder to ignore gay activists dressed as ducks picketing Fidelifacts of Greater New York to protest the company president's infamous "if one looks like a duck" comment—or the presence of cross-dressers at New York City Council hearings, gay pride marchers in New York and Los Angeles bearing signs proclaiming "Gay Is Good," high-profile photoessays in mainstream magazines such as *Life* and *Look* as well as Hollywood movies dealing with homosexual themes.[21]

As if this new visibility and militance were not challenge enough, both liberation movements called for a major transformation of American society. Unlike earlier homophile activists or liberal feminists whose initial goal was full participation without the burden of discrimination, radical feminists and gay liberationists questioned society's most basic institutions and values. The nuclear family, marriage, monogamy, traditional sex and gender roles—all were judged to be complicit in oppression. To underscore their message, liberationists supported theory with action, setting up same-sex households, nonsexist schools for children, feminist health clinics that performed abortions, shelters for battered women who needed to escape their husbands. Such actions signaled to conservatives not only the extent to which the traditional patriarchal family was being undermined, but the degree to which sex was being separated from procreation.[22]

As the 1970s progressed, traditionalists watched in increasing dismay as liberationists exercised new political clout. Gays and lesbians became an increasingly important political force, especially at the municipal level in cities such as San Francisco and New York where they were dutifully courted by politicians. In Houston, where more than two thousand delegates from every state and territory in the United States and twenty thousand guests had gathered at the 1977 International Women's Year Conference, three First Ladies—Lady Bird Johnson, Betty Ford, and Rosalynn Carter—joined with delegates to endorse feminist goals. Efforts to ratify the Equal Rights Amendment, provide federally funded child care, assure women's reproductive freedom, eliminate sex discrimination in the workplace, prevent child abuse and domestic violence, and end discrimintion against lesbians had entered the mainstream. Moreover, the presence of the three First Ladies on the platform proclaimed a message about feminists that was also boldly printed on balloons throughout the convention hall: "We Are Everywhere!"[23]

That, indeed, was the problem. So said conservatives of both sexes. Some among them could agree to the principle of equality. Others could even support specific policies such as equal pay for equal work or more women in political office. But they regarded the liberation movements of the 1960s and 1970s less as a struggle against unjust constraints than as an attack on social cohesion, moral order, family values, and personal decency.

What made the assault on containment so troubling for its defenders was that government itself seemed to have joined in the attack. With respect to foreign policy, the problem was initially Vietnam. By 1968, the effort to contain communism in that troubled land had destroyed the bipartisan foreign-policy consensus in Congress and divided the nation. The new Republican administration of Richard Nixon which assumed power in 1969 was pledged to disengagement, which supporters of the war rightly saw as an admission of defeat. Equally upsetting were other developments initiated by Nixon and his National Security Advisor Henry Kissinger. Departing from a philosophy of containment that viewed all communist regimes as identical, linked, and requiring opposition whatever the cost, the President embarked on a radically new and intellectually sophisticated policy of détente.[24]

Designed to secure friendlier relations with the Soviet Union and China and the stable balance of power globally that Nixon and Kissinger regarded as the best assurance for peace, détente was intended as a refocusing of containment, returning it to its original goal. This readjustment was not reassuring to hawks, coming as it did in tandem with the new West German policy of *Ostpolitik,* which was aimed at a closer relationship with the Soviet bloc countries of Eastern Europe and eventual reunification of the two Germanies. Objections mounted in the early 1970s as the United States made the economic and commercial concessions that détente strategists hoped would produce long-range payoffs in the form of arms control agreements.

For Americans—accustomed to think in terms of deterrence of an "evil empire" that bordered on a zero-sum game—adding carrots to the stick to reward an old foe (not yet a friend) added an unwelcome element of complexity. The policy raised uneasy questions about motivations and consequences. Did Soviet motivation stem from a genuine desire for reduced tensions or an effort to disarm the West psychologically while gaining time to improve their own position? Would the result weaken the Western alliance and diminish vigilance to contain communism? Uncomfortable with the ambiguity and complexity involved in this de-ideologizing of the Cold War and long-term resocialization of the Kremlin, skeptics were transformed into opponents by episodes in which Moscow seemed to be reverting to its old behavior. Washington, in their view, had embarked on a course of appeasement that could only lead to disaster.

Domestically, the government seemed equally culpable. The anger of conservatives found many targets: a Supreme Court that had not only destroyed racial boundaries by mandating integration, but also prohibited prayer in the public schools and struck down bans on birth control, abortion, and pornography; a federal bureaucracy that intervened in "private" relationships in the name of equality, dictating where one's children could attend school, who one's neighbors might be, and whom one's business had to employ. And there was Congress, whose willingness to implement revolution was evident to traditionalists in its passage of measures such as the Equal Rights Amendment. The White House was also deeply implicated. Not only had three First Ladies endorsed the feminist agenda adopted by delegates at the Houston Conference, but the 1980 White House Conference on Families rejected the traditional definition of family when it acknowledged multiple family forms, including unmarried heterosexual couples and even lesbian and gay couples. It was not just the heavy hand of statism on behalf of goals the conservatives did not necessarily share. It was the official sanctioning of the freedom and permissiveness associated with the 1960s. Offensive was the drug revolution, the sexual revolution, gay liberation, women's rights, students' rights, children's rights—all combined with African American, Native American, and Chicano movements into an undifferentiated assault on accustomed privileges, cherished values, and established boundaries.[25]

The centrality of boundaries, especially to many social conservatives, is evident in the reply of an angry woman who was queried as to why she opposed the Equal Rights Amendment. Her heated response had little to do with the abolition of gender distinctions in the law, which was the purpose of the amendment, and everything to do with domestic containment: "I've sat here at home and I've read the magazines, and read the newspaper, and I've watched the television news, and I never did anything. And all of a sudden, this was it. . . . There's a freedom and permissiveness that I think is wrong. But these young people who are living together, having abortions, wandering aimlessly without purpose, someday they will have to answer. There are no boundaries," she continued, ticking off a list of things that used to be considered abnormal: "homosexuality, women truck drivers, teen-age sex, and abortion." Others who shared her feelings cited additional deviations: girls playing on boys' sport teams, married women refusing to take their husband's name and using Ms., paternity leave.[26]

Many items in the lists seemed trivial from the standpoint of feminists who sought to separate sex and gender so that both men and women could enjoy more flexible social roles. But there was little that was trivial from the perspective of opponents of liberation, for whom sex and gender were inseparable and boundaries between the sexes immutable. For millions of people sexual distinction was so basic a part of selfhood, specific gender roles so firmly implanted by family of origin and so strongly reinforced by daily interaction, that gender neutrality, even in the law, was as unthinkable as same-sex marriages or the intersexual body. Consider the meaning of the following letter written immediately after Senate passage of the Equal Rights Amendment in 1972.[27]

"Today," a mother wrote to her U.S. Senator, "I am ashamed and terrified at what the future holds for my three little girls. Will my shy, sweet Tommy be drafted in six years? So modest I can't even see her undress. Oh, God! I can't stand it. I just can't bear it." To deny the relevance of sex in law was to deny life and hope. "Sex" meant the intimate privacy of shy little girls; gender neutrality meant ravaging them, stripping away the protection of innocence and thrusting them into battle. She had seen the results on television and in the press during the Vietnam War. All Americans had: ambush, rape, slaughter. It was not a linear analysis of cause and effect that Tommy's mother expressed, but the overlay of death and innocence, images that haunted rather than arguments that persuaded.

Her concern was for what her daughter's sex would mean for her as she grew up. Distinctions based on sex would allow her to ease into adulthood, unless of course she were one day confronted by a rapist, something her mother foresaw only if she became an anomaly, that is, a woman warrior in military service. "You must pass a law," the mother continued, "allowing parents to have girl children sterilized. This would be the only solution. Then women would not have to worry about or prevent pregnancy just like men. Everyone would be truly equal. Dear God in heaven, help us women." Her words conveyed anguish and protest. The fusion of ERA and sterilization shrieked symbolic danger. Tommy's mother was not being rational.

Yet she herself doubted the rationality of those whose disregard for established boundaries put her daughter at such risk. Her concern was not whether her daughter's wardrobe might eventually include a military uniform. She spoke of more compelling realities: shame and terror, and the awful assault upon her child's sexuality, the very essence of her being. She faced an absolute danger: not insecurity or carelessness or simple risk, but elemental, demonic, ultimate obliteration of self. And not just the death of the physical body, but of those values and beliefs that had shaped her Tommy in the first place. She confessed her helplessness: "Oh God . . . I just can't bear it." Moral failure and existential terror fused in the eruption of her feelings. They were very much like those of others who spoke of equality, sex, and self in ways suggesting that what was at issue was not only values and identity, but a basic understanding of what life itself was all about.

The framework of significance structuring social conservatives' beliefs reveal that for many opposing feminism—of which the amendment was a symbol—certitude about the role of sex in shaping personal identity, private obligation, family life, and social responsibility was essentially a religious conviction. Tommy's mother did not say that ERA was un-Biblical or un-Christian but that it would destroy little girls. Addressing God did not mean that she was especially religious, although she may have been, but that she had assigned to her daughter's gender—the sociocultural implications of being female—an absolute quality.

Gender, in other words, was sacred. It was a given: a biologically, physically, spiritually defined thing; an umambigious, clear, definite division of humanity into two. In an uncertain world, gender was a basic personal reminder of the orderliness of Nature. Feminists would insist that gender, like race, was a social construction; the meaning attached to sexual and racial difference was actually made by human beings, not God or Nature, and could be changed. Scholars would demonstrate that human beings who preferred love and intercourse with people of their own sex had not always been defined or identified as (fundamentally pathological) homosexuals, and that conceptions of sexuality—attitudes as to how sexual feelings should be expressed, with whom, and where—have been continually reshaped over time. Medical researchers would even point out that the dichotomous division of the sexes actually defied nature: that, biologically speaking, there are many gradations running from female to male with at least five or six sexes lying along the spectrum. But many knew better. Their certitude was unshakable that there were just two sexes—those with XX chromosomes and those with XY chromosomes—and that these biological distinctions dictated social roles and self-identities as well as opposite-sex erotic desire. Unshakable too was the conviction that preservation of fixed, bounded categories based on sexual difference was the key to moral order and social stability, and that eroding the boundaries that maintained those categories was fundamentally and irrevocably dangerous. The absolutism, certitude, and intensity associated with these beliefs made them essentially a religious conviction, irrespective of whether the individuals identified themselves as members of a religious community.[28]

Such beliefs were rejected by many Americans, especially those who had come of age during the liberation movements of the late 1960s and early 1970s. Yet these beliefs were shared by enough to suggest that domestic containment had not disappeared—quite the contrary. By the mid to late 1970s efforts were underway to revive containment internationally as well as domestically. Signs of revival were abundant: a growing anti-abortion movement, Anita Bryant's antigay campaign in Florida, the emergence of the Moral Majority, increasingly vocal opposition to détente, revisionist histories of the Vietnam War arguing that the United States could have prevailed in this "noble war" had the military been given a free hand, the popularity of *Rambo* films celebrating the perfect warrior who foiled the Russians, Viet Cong, and distant bureaucrats. If these developments seem unrelated, détente having little to do with *Roe v. Wade,* they comfortably mingled in the patriarchal discourse of the right which regarded erosion of traditional male-headed families as no less a communist objective than arms control, and Third World neutralism as abhorrent as homosexuality.[29]

The resurgence of the right in the late 1970s and 1980s reflected the determination of millions of Americans to reinvigorate containment at home as well as abroad. Conservatives' domestic legislative and judicial agenda spoke eloquently of their effort to redirect social policy: defeat of the Equal Rights Amendment, attacks on affirmative action and on family and child-care programs and policies that would help women achieve equal opportunity, denial of abortion rights, rejection of comparable worth as a legal strategy, denial of civil rights for gays and lesbians, rejection of sex education in the schools, cuts in funds for family planning for countries that gave financial support to women seeking abortions, abortion-related litmus tests for judicial appointments, and more. With respect to containment abroad, the national security agenda was equally telling: scuttling of détente, increased military spending, and defeat of the SALT II nuclear weapons treaty. Of this desire to return to the moral and intellectual simplicities of old-fashioned containment, one perceptive commentator observed, "Americans loved the Cold War too much to let it go."[30]

In Ronald Reagan they found a Cold Warrior who would both pursue the struggle and end it. To deal with the "evil empire," the President, upon taking office in 1981, reinvested containment with a moral vigor and military muscle that it had lacked since its earliest days, dramatically stepping up arms expenditures, deploying land-based intercontinental missiles, and proposing a new Strategic Defense Initiative ("Star Wars") that threatened to nullify the Soviet offensive potential. Launching a counteroffensive against Soviet expansion, the Reagan administration furnished aid to anticommunist forces in Afghanistan, Nicaragua, and other Third World trouble spots. Such policies, beginning with the demise of détente and the increased military appropriations of the later Carter years, constituted—in the judgment of some historians—a second Cold War.[31]

Although Reagan himself never promoted domestic containment with the rigor he applied to communist expansion abroad, relying often on rhetoric and posture rather than action, many of his fellow conservatives were more committed and aggressive. Waging war against gender and sexual "deviance" as though feminists and homosexuals were no less the archenemy than communists, legions on the right were nonetheless unable to prevail in an era when stemming the red tide was no longer a national obsession. The threat to nation and to democratic values, once so palpable to liberals and conservatives alike, had dissipated along with the economic dynamism that could sustain single-income dual-parent households. Still a source of certitude for some in a changing, anxiety-producing world, domestic containment was simply no longer the hegemonic discourse it had been during the first Cold War. Rather it was one among a plurality of discourses on gender and sexuality.

That as a minority discourse domestic containment endured throughout the second Cold War does not, however, explain its power and persistence *after* the demise of communism in Eastern Europe and the breakup of the Soviet Union. We need only recall the rhetoric during the presidential elections of the 1990s to be reminded that rightist fulminations against "antifamily initiatives" and homosexuals and lesbians continue to resonate with significant members of the electorate. Consider evangelist Pat Robertson's 1992 fund-raising letter in which he warned the faithful that "feminism encourages women to leave their husbands, kill their children, practice witchcraft, destroy capitalism and become lesbian." Vilification of Hillary Rodham Clinton provides another case in point. That angry demonstrators shouted "witch" when she appeared in Seattle to defend the Administration's health care plan in which she had played such a prominent role is a reminder that communities have always defined as witches those whose behavior enacts the things the community most fears—in this case, upheaval of gender hierarchy. Recent efforts to cut off public assistance for new babies of unwed welfare mothers, deny federal recognition—and spousal benefits—to same-sex married couples, and outlaw abortion are as much about domestic containment as were FBI checks on the personal sexual habits of prospective government employees during the McCarthy era. Even the rhetoric of the 1990s echoed the militarized language of an earlier era as antiabortionists targeted the American Civil Liberties Union, Planned Parenthood, and the National Organization for Women as groups to be "sought out and terminated as vermin are terminated."[32]

If domestic containment persists in the post–Cold War era, chugging along as if powered by one of those energizer batteries that never runs down, it is tempting to conclude that the explanation is one of logic. There never were inherent linkages between anticommunism on the one hand and gender and sexual containment on the other; so, logically, the latter can exist without the former—and vice versa. The logical disconnect was recognized by proponents of gay and women's rights as early as the 1950s. The most compelling demonstration of anticommunism, they argued, would be achieved not through domestic containment but by living up to the American creed and according full freedom and equality to all minorities. But simple logic, while necessary, is not sufficient to understand what was—and is—at issue for millions of Americans who continue to embrace the gender and sexual politics of containment. That additional "something" has to do with conceptions of identity, human differences, and nation.[33]

First, national identity and nationalism. We customarily think of these elusive and fluid phenomena as distinct even if closely related. Nationalism, while variously defined and chronologically located, is commonly understood to imply political expression of a group's claim for self-rule in a place peculiarly identified with its history and destiny. National identity suggests a consciousness of a common bounded community that in modern times is associated with the territorial boundaries of the nation-state. Both nationalism and national identity are invented though nonetheless real phenomena involving "imagined communities" (to use Benedict Anderson's apt phrase). We usually identify ethnicity and sometimes race as basic elements in the formation of the "imagined community," recalling old connections of family, kinship, and ethnicity embedded in claims of nationhood. Even in mature, multiethnic nation-states, the need to create a "fictive ethnicity" persists, often taking a racialized form in which ethnicity and race are equated. In the United States, individuals of various European ethnic orgins first identified as "swarthy"—Irish, Italians, Jews—gradually became "white." In the face of reason and historical evidence, these newer citizens in their turn often insisted that white Americans were the only true Americans and that other Americans (American Indians, African Americans, Hispanic Americans), some of whom had resided within U.S. territorial boundaries for far longer, were mere marginalized "others."[34]

Ethnicity and race are not the only constituent categories of nationalism and national identity. Historian George Mosse's pioneering study, *Nationalism and Sexuality,* makes clear that in the nation-states of Western Europe, especially Germany, sexuality was also fundamental. Certain sexual attitudes, desires, and practices came to be perceived as allied with nationalism, while others, which had been tolerated in earlier eras, were deemed threatening, demanding control. It is no coincidence, he argues, that with the emergence of modern nationalism in the late eighteenth and nineteenth centuries, homoeroticism had to be purged from the male bonding that was endemic in wars of liberation. Nor does he regard it as a concidence that in the nineteenth century people who preferred having sexual relations with persons of their own sex were for the first time defined as homosexuals and lesbians. Subject to new, institutionalized taxonomic discourses—medical, legal, psychological—they were marked as abnormal not merely by virtue of individual sexual acts but also in terms of physiological make-up, bodily structure, and appearance. As "outsiders" they were now those from whom "insiders" could set themselves apart. Women's sexuality also required policing—for example, via legal prohibitions on interracial marriage. Sexual intercourse with proscribed groups of men subverted the boundaries of the nation.[35]

Like sexuality, gender is embedded in nationalism and in contructions of national identities, serving as a marker of difference around which to define the "imagined community." Although gender norms within and among nations may be as varied as nations themselves, the model of the independent nation-state developed in the West within a patriarchal context. Women supported and participated in early nationalist movements. A few even disguised themselves as boys in order to fight in wars of liberation, as did Deborah Sampson who enlisted in the Fourth Massachusetts Regiment during the American Revolution. But female patriots were of use to men primarily as symbols of the nation: Germania, Britannia, and Columbia come immediately to mind. Usually depicted as idealized, static figures with a classical quality, they represented quiet strength rather than active political engagement, thereby signaling women's place in the nation these female images personifed. Since wars for nationhood are often cast as battles for manhood, the agents of national formation were and are male. In the new United States, it was arms-bearing men in the thirteen colonies who had bonded across barriers of ethnicity, religion, and class to make a revolution and write the constitution and laws that shaped a new nation in North America. The point is not that the Founding Fathers were men—we have always known that. Nor is it that men who could be antipatriarchal in their rebellion against George III could themselves be patriarchal in the inscribing of male family power in the laws of the new Republic. Rather the point is the enduring affinity of nationalism and particular forms of masculinity against other forms, and against a domesticized femininity. Even in the late twentieth century, struggles of national liberation continued to be masculine affairs with women relegated to national service as mothers, even when they have proven themselves to be skilled guerrilla fighters, as did the female soldiers of the newly formed African nation of Eritrea.[36]

Given the fusion of nationalism and particular gender forms, it is not suprising that the sharply demarcated gender roles and norms associated with nineteenth-century bourgeois society during the formative period of nation-building have become firmly if tacitly identified with the nation. Gender, like sexuality, ethnicity, and race, is always in the process of reformulation in the modern nation-state, as is national identity itself. But these formative configurations of gender, sexuality, and nationhood, while subject to change at the margins, are also remarkably persistent in their core. During moments of crisis, when the nation seems threatened, they are often reasserted, sometimes coercively, in constructions of national identity. It is at such moments that distinctions between national and alien, those who belong and those who do not, becomes most critical; without the alien there can be no national kin.

Recognizing that both nationalism and national identity are defined through the shaping of particular forms of gender and sexuality against other forms allows us to view particular historical moments with new insights, clarifying our thinking about domestic containment. The expansion of international communism in a nuclear age represented a genuine threat to the United States. That threat generated pressure to conform to founding paradigms of sexuality, gender, and nationhood as part of a new construction of national identity and purpose. Domestic containment was the result. While the Cold War gave this particular construction its distinctive coloration and, for a time, its hegemonic character, domestic containment was always subject to contestation—never more visibly and powerfully than in the late 1960s and early 1970s. Yet the linkage of anticommunism, traditional gender norms and behavior, and heterosexual marital sex, although never a logically inherent connection, continued to resonate to the very end of the Cold War for many Americans.

Why it resonated so powerfully for containment loyalists is partially explained by how they conceived of themselves and the nation. These were people like Tommy's mother for whom personal identity was rooted in the notion of fixed, unchanging human differences that were sacred. In addition, such individuals resisted the modern habit of breaking down everything, from organism to discourse, into constituent parts. Instead they exhibited a style of thinking that might be described as holistic, antimodernist, or even fundamentalist. Consequently they found it difficult to decouple sex and gender, and to separate sexuality and (hetero)sexual preference, just as earlier generations found it difficult to decouple patriotism and political ideology. Finally containment loyalists, like many people throughout history, saw the human body as a metaphor for the national body. As anthropologist Mary Douglas has explained, "the [human] body is a model which can stand for any bounded system. Its boundaries can represent any boundaries which are threatened and precarious." It is not suprising, therefore, that in the symbolic universe of such individuals, gender ambiguity, homosexuality, and communism were all assaults on the body. Nor is it an accident that communism, as historian Geoffrey Smith has noted, was likened to a disease that disfigured and destroyed, and that undetected communist cells were seen as analogous to diseased cells spreading through the human body. To prevent the spread of contagion within the national body required that strict boundaries be erected to contain communism, gender revolution, illicit sexuality, especially homosexuality, and, for some, even race mixing. (Legal barriers to interracial marriage were not overturned by the Supreme Court until 1967.) So great was the threat inherent in the blurring of boundaries that when asked to predict the consequences, respondents repeatedly answered with a single word—"chaos."[37]

If approaching the linkages associated with domestic containment in this fashion provides a fuller understanding of the dynamics of containment thought during the Cold War, we have yet to explain the gender and sexual agenda of conservative politicians and, more importantly, their constituents in the post–Cold War years. Does their persistent pressure for gender and sexual containment in the 1990s mean that pronouncements about the end of the Cold War are premature? Perhaps they are, if by the Cold War we mean a densely woven web of relationships and ideas that have sustained in the United

States both a large and lethal military structure and a particular definition of national identity. Closure, in this sense, was not achieved at the moment Berliners shook hands across the dismantled wall. There were none of the markers that Americans, remembering World War II, associate with war's end: final, punishing blows to the enemy, and victory parades. As political scientist Cynthia Enloe has observed, wars—hot and cold—are like love affairs: they don't always end neatly, but fizzle and splutter, sometimes reigniting. The persistence of domestic containment may simply be evidence of continued spluttering of a war that has not yet, in all of its manifestations, fully ended.[38]

If, however, we view domestic containment through the more capacious lens of national identity, recognizing that the production of national identity is an ongoing process, other possibilities emerge. Whether or not the Cold War has been fully terminated, its demise has surely problematized national identity once again, just as it has problematized America's international role militarily. One does not have to subscribe to Francis Fukuyama's notion of "the end of history" to acknowledge that the identity and purpose constructed for a nation functioning within the context of the Cold War will hardly suffice in a world without an "evil empire." Shattered into more than a dozen pieces, its motivating ideology discredited, the Soviet Union can no longer function as a binary opposite, as a symbol of "otherness" around which to identify, unify, and mobilize ourselves. It is no coincidence that the novelist John Updike put his character Rabbit to rest in 1990, having him lament, "Without the Cold War, what's the point of being an American?" As Updike explained, Rabbit had a concept of America "that took sharpness from contrast with Communism." "Like me, he has lived his adult life in the context of the cold war."[39]

It is hardly suprising that in the post–Cold War era efforts to define the nation are surfacing yet again in the United States. National identity, like other identities, is relational. Without an outside enemy against whom to define ourselves, conflicting groups within the nation-state seek to redefine "the people" versus "the other." Today's culture wars are a case in point. As Pat Buchanan told his fellow Republicans, they are "as critical to the kind of nation we shall be as [was] the Cold War. . . ." Evoking the passionate rhetoric and seemingly irreconcilable values associated with the epic struggle between the free world and communism, these new battles are fought over a wide range of social policy issues from abortion and multiculturalism to gays in the military. Behind this discord, many scholars have noted, is a single overriding concern: national identity. At issue are competing visions of what this nation stands for, what it has been and aspires to be, and who we are as a people. That so much of this conflict involves issues of gender and sexuality is itself a measure of just how embedded and contested these boundaries remain in contemporary constructions of American identity.[40]

Acknowledgments

Research for this chapter has been facilitated by grants from the Interdisciplinary Humanities Committee and the Academic Senate of the University of California at Santa Barbara and by the invaluable assistance of Danielle Swiontek. I am also grateful to James B. Gilbert, Linda K. Kerber, Peter J. Kuznick, and Elaine Tyler May who read an earlier draft. Members of the UCSB Cold War History Group also provided constructive criticism, especially Fred Logevall, John Talbott, and Kenneth Osgood, as did Patricia Cohen, Mary Furner, Alice O'Connor, and Ann Plane. At the University of Iowa where an early version of this chapter was presented as one of the Ida Beam lectures, I enjoyed the substantial benefits of presenting work-in-progress, especially the generosity of Kenneth Cmiel who shared his expertise on the racial dimensions of censorship of rock and roll lyrics.

Notes

1. Elaine Tyler May, *Homeward Bound: American Families in the Cold War Era* (New York: Basic Books, 1988).
2. John D'Emilio, "The Homosexual Menace: The Politics of Sexuality in Cold War America," in *Making Trouble: Essays on Gay History, Politics, and the University,* ed. John D'Emilio (New York: Routledge, 1993), 49–54; John D'Emilio, "The Bonds of Oppression: Gay Life in the 1950s," in *Sexual Politics, Sexual Communities,* ed. John D'Emilio (Chicago: University of Chicago Press, 1983), 41–46. See also Michael Kimmel, *Manhood in America: A Cultural History* (New York: Free Press, 1996), chap. 7.
3. Regina Markell Morantz, "The Scientist as Sex Crusader: Alfred C. Kinsey and American Culture," *American Quarterly* 29 (1977), 571; George Chauncey, "The Postwar Sex Crimes Panic" in *True Stories from the American Past,* ed. William Graebner (New York: McGraw Hill, 1993), 160–78; John Howard, "The Library, the Park, and the Pervert: Public Space and Homosexual Encounter in Post-World War II Atlanta," *Radical History Review* (1995), 166–87; Donna Penn, "The Sexualized Woman: The Lesbian, the Prostitute and the Containment of Female Sexuality in Postwar America," in *Not June Cleaver: Women and Gender in Postwar America,* ed. Joanne Meyerowitz (Philadelphia: Temple University Press, 1994), 358–81; John D'Emilio, "Gay Politics and Community in San Francisco since World War II," in *Hidden from History: Reclaiming the Gay and Lesbian Past,* ed. by Martin Bauml Duberman, Martha Vicinus, and George Chauncey Jr. (New York: New American Library Books, 1990), 456–73; and Estelle B. Freedman, "'Uncontrolled Desires': The

Response to the Sexual Psychopath, 1920–1960," *Journal of American History* 74 (June 1987), 83–106. Among the most notorious sexual transgressors was Christine Jorgensen whose public reception is probed in David Harley Serlin, "Christine Jorgensen and the Cold War Closet," *Radical History Review* 62 (Spring 1995), 137–65.

4. Richard J. Corber, *Homosexuality in Cold War America: Resistance and the Crisis of Masculinity* (Durham: Duke University Press, 1997), espec. Part 1.
5. For the initial statement of containment, see George Kennan, "The Sources of Soviet Conduct," *Foreign Affairs* 25 (July 1947), 566–82, and for his later thinking, *At a Century's Ending: Reflections, 1982–1995* (New York: Norton, 1996). See also Wilson D. Miscamble, *George F. Kennan and the Making of U.S. Foreign Policy* (Princeton: Princeton University Press, 1992). The development of containment policy is recounted in standard accounts of the early years of the Cold War such as: John Lewis Gaddis, *The United States and the Origins of the Cold War* (New York: Columbia University Press, 1972); Walter LaFeber, *America, Russia, and the Cold War, 1945–1967* (4th ed., New York: John Wiley, 1980); Thomas G. Patterson, *On Every Front: The Making of the Cold War* (New York: Norton, 1979). John Lewis Gaddis's more specialized study, *Strategies of Containment* (New York, Oxford University Press, 1982), traces shifts in strategy over the years. Also insightful is Deborah W. Larson, *Origins of Containment: A Psychological Explanation* (Princeton: Princeton University Press, 1985).
6. John Costello, *Virtue under Fire: How World War II Changed Our Social and Sexual Attitudes* (Boston: Little, Brown, 1985); Susan Kingsley Kent, *Making Peace: The Reconstruction of Gender in Interwar Britain* (Princeton: Princeton University Press, 1993); and Mary Louise Roberts, *Civilization without Sexes: Reconstructing Gender in Postwar France, 1917–1927* (Chicago: University of Chicago Press, 1994), 19 [quote].
7. Stephen J. Whitfield, *The Culture of the Cold War* (Baltimore: Johns Hopkins University Press, 1991), 3–4.
8. On antiradicalism in U.S. history and the first Red Scare, see John Higham, *Strangers in the Land: Patterns of American Nativism, 1860–1925* (New Brunswick, N.J.: Rutgers University Press, 1955, 1994); M. J. Heale, *American Anticommunism: Combating the Enemy Within, 1830–1970* (Baltimore: Johns Hopkins University Press, 1990); Joel Koven, *Red Hunting in the Promised Land: Anticommunism and the Making of America* (New York: Basic Books, 1994); and Kim E. Nielsen, "The Security of the Nation: Anti-Radicalism and Gender in the Red Scare of 1918–1928" (Ph.D. diss., University of Iowa, 1996). Nielsen's important dissertation revises our understanding of the first Red Scare, making abundantly clear that Bolshevism was linked not only to gender disorder but to sexual deviance and that feminism was a primary target.
9. Brett Harvey, *The Fifties: A Women's Oral History* (New York: HarperCollins, 1993).
10. Harvey, *The Fifties,* 128–46; Susan Rimby Leighow, "An 'Obligation to Participate': Married Nurses' Labor Force Participation in the 1950s," in *Not June Cleaver,* ed. Meyerowitz, 37–56.

11. Susan Hartmann, "Women's Employment and the Domestic Ideal in the Early Cold War Years," in *Not June Cleaver,* ed. Meyerowitz, 90, 96–97.
12. Harriet Hyman Alonso, "Mayhem and Moderation: Women Peace Activists During the McCarthy Era," and Dee Garrison, "'Our Skirts Gave Them Courage': The Civil Defense Peace Movement in New York City, 1955–1961," both in *Not June Cleaver,* ed. Meyerowitz, 128–50 and 201–226; Amy Swerdlow, *Women Strike For Peace: Traditional Motherhood and Radical Politics in the 1960s* (Chicago: University of Chicago Press, 1995).
13. Kathryn Nasstrom, "'Mother Love' and the Atlanta Approach to School Desegregation," paper presented at the Berkshire Conference on the History Women (Chapel Hill, N.C., June 8, 1996); Susan Lynn, "Gender and Progressive Politics: A Bridge to Social Activism in the 1960s," in *Not June Cleaver,* ed. Meyerowitz, 103–27.
14. Deborah Gerson, "'Is Family Devotion Now Subversive?' Familialism against McCarthyism,"; Ruth Feldstein, "'I Wanted the Whole World to See': Race, Gender and Constructions of Motherhood in the Death of Emmett Till"; Margaret Rose, "Gender and Civic Activism in Mexican American Barrios in California: The Community Service Organization, 1947–1962," all in *Not June Cleaver,* ed. Meyerowitz, 151–76, 263–303, and 177–200.
15. See, for example, Karen Anderson, "Women and the Politics of Race: The Little Rock School Integration Crisis, 1957–1964," paper presented at the Berkshire Conference on the History of Women (Chapel Hill, N.C., June 9, 1996).
16. On black women activists, see Vicki L. Crawford, Jacqueline Anne Rouse, and Barbara Woods, eds., *Women in the Civil Rights Movement: Trailblazers and Torchbearers, 1941–1965* (Brooklyn, N.Y.: Carlson, 1990); Jo Ann Gibson Robinson, *The Montgomery Bus Boycott and the Women Who Started It: The Memoir of Jo Ann Gibson Robinson* (Knoxville: University of Tennessee Press, 1987); Nasstrom, "'Mother Love'"; and especially Charles M. Payne, *I've Got the Light of Freedom: The Organizing Tradition and the Mississippi Freedom Struggle* (Berkeley and Los Angeles: University of California Press, 1995), chap. 9. On trade union women, see Dorothy Sue Cobble, "Recapturing Working Class Feminism: Union Women in the Postwar Era," in *Not June Cleaver,* ed. Meyerwitz, 57–83; on peace activists, see Amy Swerdlow, "The Congress of American Women: Left-Feminist Peace Politics in the Cold War," in *U.S. History as Women's History: New Feminist Essays,* ed. Linda K. Kerber, Alice Kessler-Harris, and Kathryn Kish Sklar (Chapel Hill: University of North Carolina Press, 1995), 296–312; Alonso, "Mayhem and Moderation" in *Not June Cleaver,* ed. Meyerowitz, 128–50.
17. Lillian Faderman, *Odd Girls and Twilight Lovers: A History of Lesbian Life in Twentieth-Century America* (New York: Columbia University Press, 1991), 130–40; Regina G. Kunzel, "White Neurosis, Black Pathology: Constructing Out-of-Wedlock Pregnancy in Wartime and Postwar United States," in *Not June Cleaver,* ed. Meyerwitz, 304–344; Donna Penn, "The Meanings of Lesbianism in Post-War America," *Gender & History* 3 (Summer 1991), 190–203; Daniel Gomes, "'Sissy Boys' and 'Unhappy Girls': Child-

hood Sexuality during the Cold War," paper presented at the Annual Meeting of the Popular Culture Association (Las Vegas, Nevada, March 25–28, 1996); Chauncey, "Postwar Sex Crimes Panic," in *True Stories from the American Past,* ed. William Graebner (New York: McGraw Hill, 1993), 160–78; Susan K. Cahn, "From the 'Muscle Moll' to the 'Butch' Ballplayer: 'Mannishness,' Lesbianism, and Homophobia in U.S. Women's Sports," *Feminist Studies* 19 (Summer 1992), 343–68.

18. Howard, "The Library, the Park, and the Pervert," 166–87; Rickie Solinger, "'Extreme Danger': Women Abortionists and Their Clients before *Roe v. Wade,*" in *Not June Cleaver,* ed. Meyerwitz, 335–57.

19. *Rock and Roll Personality Parade* (London: New Musical Express, 1957), 5 [Haley quote]; Linda Martin and Kerry Segrave, *Anti-Rock: The Opposition to Rock and Roll* (Hamden, Conn.: Archon Books, 1988); John A. Jackson, *Big Beat Heat: Alan Freed and the Early Years of Rock & Roll* (New York: Schirmer Books, 1991). Jackson notes that there was also opposition from upper- and middle-class blacks who objected to lyrics about sexual promiscuity and other deviant behavior because it reinforced negative racial sterotypes. Policing of rock and roll did not begin until 1954 when white audiences became involved. The Supreme Court's ruling that year in the *Brown* decision may also have generated fears about race-mixing, especially among teenagers, a concern uppermost in the minds of many southern opponents of school integration. On efforts of the Los Angeles police force to prevent race-mixing, see George Lipsitz, "Land of a Thousand Dances: Youth, Minorities, and the Rise of Rock and Roll," in *Recasting America: Culture and Politics in the Age of Cold War,* ed. Lary May (Chicago: University of Chicago Press, 1989), 267–84, 274 [quote].

20. Joanne Meyerowitz, "Beyond the Feminine Mystique: A Reassessment of Postwar Mass Culture, 1946–1958," and Wini Breines, "The 'Other' Fifties: Beats and Bad Girls," both in *Not June Cleaver,* ed. Meyerowitz, 231–62, 382–408; Ruth Rosen, "The Female Generation Gap: Daughters of the Fifties and the Origins of Contemporary American Feminism," in *U.S. History as Women's History,* eds. Kerber, Kessler-Harris, and Sklar, 313–34; Barbara Ehrenreich, *Hearts of Men: American Dreams and the Flight from Commitment* (Garden City, N.Y.: Anchor Books, 1983); John D'Emilio and Estelle B. Freedman, *Intimate Matters: A History of Sexuality in America* (New York: Harper & Row, 1988), 302–5; Allan Berube, "Marching to a Different Drummer: Lesbian and Gay GIs in World War II," in *Hidden from History,* eds. Duberman, Vicinus, and Chauncey, 383–94; Faderman, *Odd Girls and Twilight Lovers,* 139–87; D'Emilio, "Gay Politics and Community in San Francisco" in *Hidden from History,* 456–73; Lewis A. Erenberg, "Things to Come: Swing Bands, Bebop, and the Rise of a Postwar Jazz Scene" in *Recasting America,* ed. Lary May, 221–45. For other evidence of contestation in the early 1960s, see Blanche Linden-Ward and Carol Hurd Green, *American Women in the 1960s: Changing the Future* (New York: Twayne Publishers, 1993).

The enduring linkage within the right of patriotism and traditional gender norms is exemplified not only in the 1950s but in the pre–World War II campaign for isolationism. That linkage must be understood not only in terms of patriarchy, but also in terms of nationalism, as I explain subsequently. It can be embraced by women as well as men. See Laura McEnaney, "He-Men and Christian Mothers: The America First Movement and the Gendered Meanings of Patriotism and Isolationism," *Diplomatic History* 18 (Winter 1994), 47–57; also Glen Jeansonne, *Women of the Far Right: The Mothers' Movement and World War II* (Chicago: University of Chicago Press, 1996).

21. Neil Miller, *Out of the Past: Gay and Lesbian History from 1869 to the Present* (New York: Vintage Books, 1995), 378–84; D'Emilio and Freedman, *Intimate Matters*, 318–20. The president of Fidelifacts, a credit agency, was quoted as saying, "If one looks like a duck, walks like a duck, associates only with ducks, and quacks like a duck, he is probably a duck."
22. Miller, *Out of the Past*, 369–70; Andrew Hum, "The Personal Politics of Lesbian and Gay Liberation," *Social Policy* 11 (October/September 1980), 43; Margaret Cruikshank, *The Gay and Lesbian Liberation Movement* (New York: Routledge, Chapman & Hall, 1992), 57–107; D'Emilio and Freedman, *Intimate Matters*, 321–23; Faderman, *Odd Girls and Twilight Lovers*, 188–214; Alice Echols, *Daring To Be Bad: Radical Feminism in America* (Minneapolis: University of Minnesota Press, 1989).
23. Arthur Goldberg was the first New York gubernatorial candidate to announce his support of gay rights in his race against Nelson Rockefeller in 1970. See Miller, *Out of the Past*, 379–82, 395–99; also Faderman, *Odd Girls and Twilight Lovers*, 199–201. On the Houston Conference, see Carolyn Bird and the Members and Staff of the National Commission on the Observance of International Women's Year, *What Women Want: From the Official Report to the President, the Congress, and the People of the United States* (New York: Simon & Schuster, 1979), 83–178. Other items on the Houston women's agenda included the development of women's studies programs, civil rights for disabled women as well as women of color and lesbians, greater participation of women in the making of foreign policy and in the media, and the strengthening of rape laws.
24. Discussion of détente in this and the following paragraphs relies on Gordon A. Craig and Alexander L. George, *Force and Statecraft: Diplomatic Problems of Our Time* (3d ed., New York: Oxford University Press, 1995), chap. 10; and H. W. Brands, *The Devil We Knew: Americans and the Cold War* (New York: Oxford University Press, 1993). For more detailed treatment, see Kjell Goldmann, *Change and Stablity in Foreign Policy: The Problems and Possibilities of Détente* (Princeton, N.J.: Princeton University Press, 1988). See also Henry A. Kissinger, *Diplomacy* (New York: Simon & Schuster, 1994), chap. 29.
25. The negative response to feminism has been much elaborated in my other writings. See, for example, Donald G. Mathews and Jane Sherron De Hart, *Sex, Gender, and the Politics of ERA: A State and the Nation* (New York: Oxford University Press, 1990), espec. chaps. 2, 6–8. See also Rebecca Klatch, *Women of the New Right* (Philadelphia: Temple University Press, 1987); Pamela Johnston Conover and Virginia Gray, *Feminism and the New Right: Conflict over the American Family* (New York:

Praeger, 1983); Rosalind Pollack Petchesky, "Antiabortion, Antifeminism, and the Rise of the New Right," *Feminist Studies* 7 (1981), 206–46; and Jerome L. Himmelstein, *To the Right: The Transformation of American Conservatism* (Berkeley: University of California Press, 1990), espec. chap. 4. For neoconservatives the threat is not only to family and personal responsibility, but to the work ethic and, therefore, capitalism itself. See, for example, Daniel Bell, *The Cultural Contradictions of Capitalism* (New York: Basic Books, 1975).

26. ERA opponent interviews, by prior agreement, are cited without the name of the interviewee. This interview occurred in 1977 and is included in the listing of interviews in Mathews and De Hart, *Sex, Gender, and the Politics of ERA,* 272–75.
27. Mrs. Alonzo Mayfield to Senator Samuel J. Ervin Jr. (March 1972), Papers of Senator Samuel James Ervin Jr., Southern Historical Collection, University of North Carolina, Chapel Hill, N.C. Analysis of the centrality of gender in this letter follows closely the text of my "Gender on the Right: Meanings Behind the Existential Scream," *Gender & History* 3 (Autumn 1991), 246–67; also Mathews and De Hart, *Sex, Gender, and the Politics of ERA,* 161–66. The term "intersex" or "intersexual" is used medically as a catch-all for three major subgroups with some mixture of male and female characteristics: true hermaphrodites who possess one testis and one ovary; male pseudo-hermaphrodites who have testes and some aspects of female genitalia but no ovaries; and female pseudo-hermaphrodites, who have ovaries and no testes but possess some aspects of male genitalia. Contemporary medical practice is to treat intersexuality, which is estimated to occur in about 4 percent of births, as disease, intervening hormonally and surgically so that these individuals can be categorized as male or female. That practice, which reflects society's insistence on maintaining distinctions betwen the sexes, has recently been called into question in the medical community. See Anne Fausto-Sterling, "How Many Sexes Are There?" *New York Times* (March 12, 1993), A29.
28. For futher discussion of this fusing of cultural patterns and the sacred cosmos in a style of thinking that rejects the flexibility of cultural forms, resists the moral implications of complexity and ambiguity, and clings to binary concepts, see Mathews and De Hart, *Sex, Gender, and the Politics of ERA,* 168–69; and espec. Donald G. Mathews, "Spiritual Warfare": Cultural Fundamentalism and the Equal Rights Amendment, *Religion and American Life* 3 (Summer 1993), 129–54. Gender and sexuality are not the only matters perceived in this fashion. Antipathy to abstract expressionist art in this period suggested a similiar response that I labeled "cultural fundamentalism" in my "Art and Politics in Cold War America," *American Historical Review* 81 (October 1976), 762–87.
29. Dallas A. Blanchard, *The Anti-Abortion Movement and the Rise of the Religious Right: From Polite to Fiery Protest* (New York: Twayne, 1994); Michael Lienesch, *Redeeming America: Piety and Politics in the New Christian Right* (Chapel Hill: University of North Carolina Press, 1993); Freedman and D'Emilio, *Intimate Matters: A History of Sexuality in America* (New York: Harper and Row, 1988), chap. 15; Susan Jeffords,

The Remasculinization of America: Gender and the Vietnam War (Bloomington and Indianapolis: Indiana University Press, 1989); Craig and George, *Force and Statecraft,* chap. 10. Revisionist analyses of Vietnam include: Guenter Lewy, *America in Vietnam* (New York: Oxford University Press, 1978); Norman Podhoretz, *Why We Were in Vietnam* (New York: Simon & Schuster, 1982); Harry G. Summers Jr., *On Strategy: A Critical Analysis of the Vietnam War* (Novato, Ca.: Presidio Press, 1982); Timothy J. Lomperis, *The War Everyone Lost–and Won: America's Intervention in Viet Nam's Twin Struggles* (Baton Rouge: Louisana State University Press, 1984); Bruce Palmer Jr., *The 25-Year War: America's Military Role in Vietnam* (Lexington: University Press of Kentucky, 1984).

The commonality among these various developments can also be explained conceptually in terms of the nation as a "family" and what George Lakoff calls the "strict father" model of parenting. A "strict father" defends his "family" from external enemies and upholds the strict, clear moral boundaries that are necessary to maintaining moral order. See George Lakoff, *Moral Politics: What Conservatives Know That Liberals Don't* (Chicago: University of Chicago Press, 1996).

30. This agenda is embraced with particular intensity by the new Religious Right that has its deepest roots among white evangelical and fundamentalist Christians. For a balanced and informative treatment of the ideas that inform their political agenda, see Lienesch, *Redeeming America.* For a historical treatment of fundamentalists' views on gender, see Margaret Lamberts Bendroth, *Fundamentalism and Gender: 1875 to the Present* (New Haven: Yale University Press, 1993). On the socioeconomic base of contemporary antifeminism, see Jerome L. Himmelstein, "The Social Basis of Antifeminism: Religious Networks and Culture," *Journal for the Scientific Study of Religion* 25 (March 1986), 1–15. See also Jerome L. Himmelstein and James A. McRae Jr., "Social Issues and Socioeconomic Status," *Public Opinion Quarterly* 52 (Winter 1988), 492–512. On antiabortion activists, see Kristin Luker, *Abortion and the Politics of Motherhood* (Berkeley and Los Angeles: University of California Press, 1984); also Rosalind Pollack Petchesky, *Abortion and Women's Choice: The State Sexuality and Reproductive Freedom* (Boston: Northeastern University Press, 1984, 1990), espec. chaps. 7–8. On the right's impact on policy issues relating to women and equality in the 1980s, see my "Equality Challenged: Equal Rights and Sexual Difference," *Journal of Policy History* 6 (Spring 1994), 40–72.

On opposition to détente and Salt II, see Brands, *The Devil We Knew,* 152–62, 163 [quote]. Kissinger argued that conservatives were so committed to viewing the conflict with communism as ideological that they were unable to view it in geopolitical terms as required by détente. Hardline conservatives (not neoconservatives), he maintained, were such "moral absolutists" that they distrusted any negotiation with the Soviet Union, "viewing compromise as retreat." See Kissinger, *Diplomacy,* 742–45.

31. Brands, *The Devil We Knew,* 163–86; Craig and George, *Force and Statecraft,* 129–31. Kissinger's *Diplomacy* is also well worth reading for his assessment of Reagan's "astonishing performance" in his conduct of the Cold War.

32. Robertson's statement, made in a fund-raising letter, is quoted in Jason DeParle, "A Fundamental Problem," *New York Times Magazine* (July 14, 1996), 24. On the Seattle incident involving Hillary Clinton, see the *New York Times* (January 1, 1995), A11. The characterization of pro-choice groups as "vermin" was a section of a letter to Congress quoted in part in the *New York Times* (January 1), A8. This militarized rhetoric is part of the larger phenomenon of militarization of American society superbly treated in Michael S. Sherry, *In the Shadow of War: The United States Since the 1930s* (New Haven: Yale University Press, 1995).
33. Exploring arguments that anticommunism did not dictate domestic containment but rather required extension of rights to all minorities, Joanne Meyerowitz has also made the point that there is no inherent linkage between anticommunism and gender and sexual containment. See her "Gender, Sex, and the Cold War Language of Reform," paper presented at the 1994 Landmarks Conference on the Cold War and American Culture (Washington, D.C., March 17–19, 1994). It is worth noting that Americans (48 percent) remain firmly attached to traditional gender roles—more so than are their counterparts in many other countries. In an international poll of adults in twenty-two nations conducted in 1995 by the Gallup Organization, only in Hungary and Chile did the traditional family with breadwinner father and a homemaker mother receive greater support. See *New York Times* (March 27, 1996), A12.
34. Benedict Anderson, *Imagined Communities: Reflection on the Origin and Spread of Nationalism* (London: Verso, 1983), 49, 66. Etienne Balibar, "The Nation Form: History and Ideology," trans. Immanuel Wallerstein and Chris Turner, *Review, Fernand Braudel Center* 13 (Summer 1990), 322–61; also Etienne Balibar and Immanuel Wallerstein, *Race, Nation, Class: Ambiguous Identities* (London: Verso, 1991). Balibar's use of the adjective "fictive" is not meant to imply that enthnicity is not grounded in lived experience or that it does not have real consequences. Rather the emphasis is on the social construction over time of what seems to be a "natural" and timeless category. For an important statement on ethnicity as a social construction by leading U.S. scholars, see Kathleen Neils Conzen, David A. Gerber, Ewa Morawska, George E. Pozetta, and Rudolph J. Vecoli, "The Invention of Ethnicity: A Perspective from the U.S.A.," *Journal of American Ethnic History* 12 (Fall 1992), 4–32. For the process by which various European immigrant groups underwent a kind of transethnic, racial homogenization that enabled them to claim whiteness, see David Roediger, "Whiteness and Ethnicity in the History of 'White Ethnic' in the United States," in Roediger, *Towards the Abolition of Whiteness: Essays on Race, Politics, and Working Class History* (London: Verso, 1994), and *The Wages of Whiteness: Race and the Making of the American Working Class* (New York: Verso, 1991); also Noel Ignatiev, *How the Irish Became White* (New York: Routledge, 1995), and Arnold R. Hirsch, *Making the Second Ghetto: Race and Housing in Chicago, 1940–1960* (New York: Cambridge University Press, 1983). A superb analysis of the process by which Euro-Americans in the eighteenth century established a sense of national identity that imagined white Americans as the only true Americans is provided by Carroll

Smith-Rosenberg, "Captured Subjects/Savage Others: Violently Engendering the New Republic," *Gender & History* 5 (Summer 1993), 177–95.

35. George L. Mosse, *Nationalism and Sexuality: Respectability and Abnormal Sexuality in Modern Europe* (New York: Howard Fertig, 1985). See also *Nationalisms and Sexualities,* ed. Andrew Parker, Mary Russo, Doris Sommer, and Patricia Yaeger (New York: Routledge, 1992), and Anne McClintock, *Imperial Leather: Race, Gender, and Sexuality in the Colonial Context* (London: Routledge, 1994). The homosocial affinities of nationalism are not meant to obscure the greater permissiveness of heterosexual standards of sexual expression in the age of the American Revolution as compared with the nineteenth century.

36. For new scholarship that explores gender and nationalism in an international context, see the special issue of *Gender & History* 5 (Summer 1993); and Anne McClintock, "No Longer in a Future Heaven: Nationalism, Gender, and Race," in *Becoming National,* ed. Geoff Eley and Ronald Grigor Suny (New York: Oxford University Press, 1996), 260–84. On the intersection of gender and race in the production of the new American nation, see Carroll Smith-Rosenberg, "Dis-Covering the Subject of the 'Great Constitutional Discussion,' 1786–1789," *Journal of American History* 79 (December 1992), 841–73. Apart from Smith-Rosenberg, U.S. historians have been far less active with respect to this topic than have literary scholars. See, for example, *Subjects and Citizens: Nation, Race, and Gender from Oroonoko to Anita Hill,* ed. Michael Moon and Cathy N. Davison (Durham, N.C.: Duke University Press, 1995). On women as symbols of the nation, see Mosse, *Nationalism and Sexuality,* chap. 5. Perhaps the one exception is Marianne, who, as Sara Kimble informed me, played a much more active role as the symbol of French nationalism. See Sara L. Kimble, "Deciphering Marianne: Gender in French Republican Caricature, 1870–1871" (MA thesis, University of Iowa, 1994). The women who fought as guerillas in Eritrea are by no means unique in being told by the male political leadership to go home and be good mothers. In response to similar admonitions, a Salvadoran woman who has fought with the Farabundo Martí National Liberation Front, the FMLN, after dutifully handing over her gun, had her IUD removed as one of her first postwar acts. See Cynthia Enloe, *The Morning After: Sexual Politics at the End of the Cold War* (Berkeley and Los Angeles: University of California Press, 1993), 1.

37. Mary Douglas, *Purity and Danger: An Analysis of Concepts of Pollution and Taboo* (London: Routledge and Kegan Paul, 1966), 115. Geoffrey S. Smith, "National Security and Personal Isolation: Sex, Gender, and Disease in the Cold-War United States," *International History Review* 14 (May 1991), 307–37. That the threat posed by blurring of boundaries was thought to be catastrophic was evident in answers to an open-ended question posed in an opinion survey of North Carolina women conducted by the Institute for Research in the Social Sciences at the University of North Carolina at Chapel Hill in 1978 with which I was involved. The question—What do you think will happen if ERA is ratified?—was repeatedly answered with one word: "chaos."

That segregationists would include race-mixing in this blurring of boundaries is to be expected, especially since notions of kinship, blood, and family projected outward construct our notions of nationality and citizenship. See David Schneider's discussion of how white Anglo-Saxon Protestants construct notions of kinship in his *American Kinship: A Cultural Account* (Chicago: University of Chicago Press, 1968), and, more important, Ramon A. Gutierrez, "Decolonizing the Body: Kinship and Nation," *American Archivist* 57 (Winter 1994), 86–99.

38. Enloe, *The Morning After,* 2.
39. John Updike, *Rabbit at Rest* (New York, 1990), 442–43; John Updike, "Why Rabbit Had to Go," *New York Times Book Review* (August 5, 1990), 27.
40. The problematizing of national identity in the wake of the demise of communism in Europe may also have a demographic dimension. The Cold War made it imperative that the United States drop its racist immigration restriction laws and offer nonwhite Third World nations an alternative to communism, as Secretary of State Dean Rusk argued in 1965. But the greater ethnic and racial diversity associated with the huge influx of Asian and Latino immigrants admitted under the new immigration laws of 1965 and 1991 and the related phenomenon of identity politics has complicated maintenance of that "fictive" ethnicity of whiteness so basic to American national identity in the past. On the culture wars and national identity, see Pat Buchanan, "The Election Is about Who We Are," speech delivered at the Republican National Convention, Houston, Texas (August 17, 1992), in *Vital Speeches of the Day.* Typical of scholarly studies is the assessment contained in James Davidson Hunter, *Culture Wars: The Struggle to Define America* (New York: Basic Books, 1991), 50. While acknowledging that the culture wars are about defining America and that many rage around issues involving gender and sexuality, students of contemporary cultural conflict in the United States have not explored the linkages among sexuality, gender, and national identity—a task I attempt in forthcoming studies tentatively titled *Litigating Equality* (Chicago: University of Chicago Press, 2001) and *Defining America* (2002).

CHAPTER 7

"Cold War Culture" Doesn't Say It All

Peter Filene

There's something irresistibly convenient about the concept "Cold War culture."[1] A mating of foreign and domestic; a package you can carry in one hand, leaving the other free for broad gestures. It's not as baggy as that once-fashionable idea, "national character." Nor is it as empty as "postwar America." Understandably, then, historians have been deploying it in all sorts of fields: race, gender, film, to name a few. As Geoffrey Smith has remarked: "Only now have historians begun to sense that this conflict [the Cold War] became a rationale for nearly everything in American life and culture from the GI Bill and college loans, to Eisenhower's interstate highway system (to escape nuclear attack), to civil rights for African Americans, to Dan Quayle himself."[2]

Most historians would not extend that rationale beyond the mid-60s, by which time—to quote Stephen Whitfield—the culture of the Cold War had "decomposed."[3] They agree, however, that American postwar culture was infused by the fear of communism abroad and at home.

A few years ago I encountered a disconcerting flaw in this all-purpose interpretation. In my undergraduate survey course, I asked the ninety students to interview a grandparent about his or her memories of postwar life. (Most of the students, by the way, were white, middle-class Protestants from small towns and rural areas in North Carolina.) As a guideline for their collective oral history, I quoted our textbook: "While troubles abroad deeply disturbed Americans, domestic witch-hunting in the late forties and early fifties" had produced a "national paranoia over communism," accompanied by "widespread anxieties."[4] To what extent did this interpretation match the particular experiences of their grandparents?

A week later, the students came back with surprising reports. Their grandparents had not been deeply disturbed by communist aggression in Eastern Europe and Korea nor had they felt "paranoid" or even particularly anxious about McCarthyism. With few exceptions, the respondents recalled having been concerned back then (in their twenties) with holding a job, paying the rent, and raising young children. As I polled each student and scribbled his or her evidence on the blackboard, I felt the consternation spreading through the room. Was the textbook wrong? But it told the same story they had learned in other courses, a story that had made unquestionable sense, at least until now. Yet how could there be two contradictory versions of the same time period? It was one of those classroom moments I treasure, when my students and I are suddenly stymied, caught up in a quest for new answers.

This chapter is my continuation of that quest. The concept of "Cold War culture" offers a significant reframing of American history after 1945. Whereas previous historiography had generally divorced foreign events from domestic developments, this recent perspective has been discovering intriguing relationships between them. Inconsistent evidence prompts me to issue a partial dissent. I'm going to argue that the Cold War was fought primarily at an elite level. It pervaded and shaped the experience of ordinary Americans far less than historians would have us believe. Although government leaders, social-science experts, and media commentators set the terms of public discourse—and also of public policy—most citizens to a surprising degree defined their world in personal terms.

This discrepancy between the elite and the majority emerges starkly in opinion polls about the communist threat. The concerns in Washington were not the concerns on Main Street. The gap emerges more intricately in the realm of gender. Policymakers invoked traditional gender stereotypes: "tough" anticommunism, along with a nuclear family secured by the male breadwinner and the female homemaker. To a striking degree, on the other hand, men and women of the fifties deviated from these prescriptions, and the young radicals of the sixties openly defied them. In the world of popular entertainment, meanwhile, we find a similar discrepancy. Hollywood producers made movies and television shows directly invoking the Red threat, only to find that audiences preferred unpropagandistic make-believe. By the end of the chapter, I hope not to have dismantled the concept of "Cold War culture," but to have complicated it.

Public Opinion

Whatever else we may say about the decade following V-J Day, surely it was an era of communist menace abroad and at home. *The Great Fear, Nightmare in Red, The Dark Ages*—these are some of the lurid labels that historians have devised, and of course there's Arthur Miller's *The Crucible.*[5] Even scholars who cheerfully portray *The Proud Decades* or *American High* make an exception for the witch-hunts that uncovered at best a few spies while spreading a pall of fear and conformity among the rest of the population.[6]

How do we reconcile this description with the indifference to the Red menace among my students' grandparents? Perhaps in the course of forty years they had forgotten what they thought about world events.[7] Or perhaps, like survivors of the Great Depression, from the safety of hindsight they were casting the experience in positive terms.[8] Or maybe for some reason North Carolina escaped the "nightmare" that tormented the rest of the nation.[9] Each of these hypotheses holds a bit of truth, but ultimately they don't explain away the inconsistency. For when we look at public opinion polls during the postwar decade, it becomes clear that the majority of Americans were far less obsessed with communism than historians have been.

Consider, to begin with, this fact of daily behavior: Although 80 percent of adults in the 1950s received a daily newspaper, the average metropolitan reader spent four minutes on the important news and forty minutes on the comics, sports, and other entertaining features beyond the front page.[10] National politics—the stuff of traditional history—earned (and still earns) little attention from the man on the street, and foreign affairs even less. One can best visualize American public opinion as three concentric circles: a core of policymakers, numbering a few thousand; a ring of college-educated, well-to-do people who devote close attention to international events and who serve as opinion leaders, numbering two to three million in the 1950s; and an outer ring including everyone else, the 90 percent who pay fitful attention, if any, to news of the world beyond the shoreline.[11]

The threat of war will penetrate this indifference. As Nazi and Japanese aggression mounted during 1935–1941, so did the proportion of Americans who named foreign problems as "the most important facing the American people," reaching 81 percent on the eve of Pearl Harbor. But once the United States entered the war, they resumed their parochial outlook. Inflation, food shortages, and labor union strife became the main concerns. Shortly after V-J Day, only 7 percent gave top rank to foreign problems.[12]

Victory didn't produce peace. By 1947, it had evolved into a Soviet-American belligerence that was neither war nor peace—a "cold war." By 1949, the Soviets had exploded an atomic bomb and the communists had seized power in China. Newsreels and weekly magazines drew maps showing a Red tide spilling down toward Europe and Southeast Asia. If public opinion responds to the possibility rather than the actuality of international danger, then one would expect a surge in foreign-policy attention. That is indeed what happened. During the late 1940s and 1950s, the threat of war ranked first among the problems facing the American people.[13] But unemployment and the high cost of living came close behind, suggesting that concern with foreign policy was competing with matters closer to home. Truman's advisers warned him that if he was to win congressional approval for anticommunist aid to Greece and Turkey, he would have to "scare the hell out of the country."[14] The Truman Doctrine served as "a form of shock therapy," to quote one historian. Yet its effects wore off quickly. Within a few months, public concern with foreign affairs dropped almost in half.[15]

Meanwhile, the House Committee on Un-American Activities (HUAC) intensified its crusade against the communist subversives allegedly at work within the federal government, labor unions, schools, and other arenas of American life. In 1950, Joseph McCarthy became the crusade's self-appointed leader. The Red menace had crossed the shoreline. Like polio, it was an invisible, capricious virus that could paralyze the body politic without warning and without cure.[16] A district judge refused bail to five aliens while authorities decided whether to deport them. "I am not going to turn these people loose if they are communists," the judge explained, "any more than I would turn loose a deadly germ in this community." As moviegoers saw in *The Invasion of the Body Snatchers* (1956), the enemy lurked around us inside human-looking forms. "With the source of the evil so elusive and so immune to risk-free retaliation," historian Stephen Whitfield declared, "American culture was politicized."[17] Cold War and culture merged into a phrase.

The fact is, however, that most Americans had other worries on their minds. Historians may cast McCarthy as the leading actor in this drama. But in 1953, at the height of his notoriety, 29 percent of the public said they had an unfavorable opinion of the senator, while 21 percent had no opinion at all.[18] Even more astonishingly, fewer than one out of five thought communism in the United States was the country's or the government's "most important problem."[19] When asked "what is the main problem facing your section of the country?" southerners named race; westerners named conservation; easterners and midwesterners named prices.[20]

These were, one might say, the front-page-of-the-newspaper questions, raising the kinds of issues to which the average citizen devoted four minutes before (or after) turning to the materials that really interested him or her. We shouldn't be surprised, then—although I continue to be astounded—by the response to the more personalized question, "What is your biggest worry these days, the thing that disturbs you most?" Only 1 percent mentioned communism in the United States. The main worry for almost half of Americans in 1953 was the high cost of living.[21]

In short, there were two realms of experience, elite and ordinary, which coexisted but rarely intersected, much like the new interstate highways and the old two-lane roads. It was a heterogeneous culturescape. Accordingly, we need to multiply "the national mood" into two moods or more.

Gender

Although most Americans ignored the Cold War most of the time, they couldn't escape its effects on their daily lives. Defense jobs proliferated on the Pacific coast, while federal research and development funds streamed into universities. A wave of anticommunist investigations destroyed dissidents' careers or forced them to camouflage their ideas in Cold War language.[22] Civil defense drills sent children under their desks, and the Korean War put several hundred thousand young men into uniform. To this extent, the Soviet-American conflict undeniably infiltrated American life.

The concept of a "Cold War culture" becomes more interesting and debatable, though, when one applies it to an arena that seems to have nothing to do with foreign affairs: namely, gender and family. Until recently, these two fields remained as separate as issues of *Diplomatic History* and *Signs* on the library shelves. Yet this very separation has gendered implications. As Amy Kaplan astutely remarked, historians have worked from the binary assumption that world affairs are masculine activity while domestic affairs are feminine.[23] When one examines the Cold War through the lens of gender, old categories dissolve and new meanings emerge.

Consider, for example, the language of policymakers. "It was physical hardihood that helped Americans in two great world wars to defeat strong and tenacious foes," President Kennedy told readers of *Sports Illustrated* in 1962. "And today . . . in the jungles of Asia and on the borders of Europe, a new group of vigorous young Americans helps maintain the peace of the world and our security as a nation." In private, President Johnson was more vulgar. When he heard that a member of his administration was becoming a dove on Vietnam, Johnson muttered: "Hell, he has to squat to piss."[24] To be a man was to be not feminine. Hard, not soft. Hard-headed, hard-nosed, hard-assed.

Macho rhetoric was nothing new in American public life. But the Cold War endowed it for the elite with extra fervor and currency: "a cult of toughness."[25] After Soviet athletes trounced Americans in the summer Olympics of 1956 and 1960, a *Saturday Evening Post* writer called for "toughening our soft generation."[26] Arthur Schlesinger contrasted "the new virility" of liberalism with the "political sterility" of left-leaning ideologues like Henry Wallace.[27] In the world of pulp fiction, Mickey Spillane was more graphic. "I killed more people tonight than I have fingers on my hands," boasted detective Mike Hammer in *One Lonely Night* (1951). "I shot them in cold blood and enjoyed every minute of it. . . . They were commies. . . . They were red sons-of-bitches who should have died long ago. . . . They never thought there were people like us in their country. They figured us all to be soft as horse manure and just as stupid." Of the top ten bestselling novels in the 1950s, six were by Spillane.[28]

If Cold Warriors were manly, their opponents must be effeminate. McCarthy sneered at the "striped pants boys in the State Department," while Everett Dirksen promised to fire those "lavender lads." In case anyone missed the point, their fellow senator from Nebraska explained: "You can't . . . separate homosexuals from subversives."[29] There was a certain truth to this allegation, because gay men in that homophobic era were vulnerable to blackmail by communist agents. Otherwise it was a form of scapegoating—a "lavender scare"—fueled by men's need to define their sexual identity in opposition to "the other." During the early 1950s, an average of sixty homosexuals were fired from federal jobs each month as security risks.[30]

Cultural spokesmen applied this rigid gender system (and Cold War language) not only to the political homefront, but to the home itself. Amid the tensions of the postwar world, experts in various fields endorsed the traditional nuclear family as the cornerstone of the nation's security. The assignment for women was clear. Whatever sexual pleasures and employment ambitions they might have enjoyed during the World War, these were to be contained within marriage. Women were supposed to enjoy domesticity as their career, using their energy and talents for the betterment of their families and, via volunteer work, their communities.[31] The Federal Civil Defense Administration, for example, launched a publicity campaign for families to set up home bomb shelters on the model of Grandma's pantry: canned goods, soap, candles, a first-aid kit, a portable radio. In so doing, a government spokeswoman said, "a mother must calm the fears of her child. Make a game out of it: Playing Civil Defense."[32] The cultural expectations for men were equally clear. "Virtually the only way to be a real man in our society," asserted a leading sociologist, "is to have an adequate job and earn a living."[33] Father, the manly breadwinner; mother, the womanly nurturer; and four children: it seemed a revival of the Victorian family model.

In two respects these prescriptions by the elite didn't end up dictating the experiences of the people at whom they were aimed. Or speaking more formally, gender roles didn't match gender identities. First, there were serious inconsistencies that cut across the model like stress lines in the wall of a house. Women could contain themselves within the housewifely role, but there was chafing along the way. As one said, "I feel I am doing exactly as I am fitted—with an occasional spurt of independence growing less all the time."[34] While many housewives had trouble with their domesticity script, others rewrote it. In the 1950s, ten million women—including an unprecedented proportion of middle-class wives—were employed at least part-time. They said they were working to boost the family income, not to satisfy personal ambitions. But given the fact that behavior shapes attitudes, female employment signaled a flaw in the *Good Housekeeping* model.[35]

Men were contending, meanwhile, with inconsistencies in their cultural assignment. Besides earning the family income, in this age of "togetherness," the breadwinner must also play with wife and kids on the home team. A real man was supposed to wear the pants in the family, but also grill the T-bone steak in the backyard. Measured against the Cold War cult of toughness, such men seemed tame. There was uncomfortably little distance between the man in the gray flannel suit and the striped pants boys. "Being a real father is not 'sissy' business," a psychiatrist stated in *Parents' Magazine.*[36] His insistence betrayed the fact that a lot of fathers believed otherwise. In *Rebel without a Cause,* James Dean's father literally wore an apron as he bent to his wife's nagging demands and failed to stand up for his son. The popular press fretted about "the domestication of the American male."[37]

Postwar opinion leaders were broadcasting a mixed message to men: toughness, but also teamwork; domesticity, but not too much of it. To escape the tender trap, many men resorted to mediated fantasy. Throughout the '50s and well into the '60s, John Wayne headed the popularity polls as the American man whom American men admired most. Eight of the top ten television shows were "horse operas," and in 1958 alone Hollywood produced fifty-four cowboy movies.[38] Baseball may have been the national pastime, but pro football was the bonebreaking sport that most men watched on television in their pine-paneled family rooms or in their local bar.[39] Boys crowded movie houses to see Walt Disney's young-man-coming-of-age adventure films. In fact, maybe not only boys. According to the *New York Times* movie critic, *Treasure Island* would set men "to daydreaming about excitements which are denied fulfillment by the inhibitions of advancing age."[40]

Hugh Hefner also had masculine excitements in mind when he launched

Playboy magazine in 1953 and quickly won a large readership. As he announced in the premier issue, *Playboy* offered men "a little diversion from the anxieties of the Atomic Age." Yet as subsequent issues made unmistakably clear, the diversion was from domesticity. The hardworking breadwinner who drove home in his station wagon to kiss his wife and help his kids with their homework could read, in *Playboy*'s glossy pages, about fancy sports cars, sports jackets, stereos, and no-strings-attached sexual pleasure.[41]

Women were pushing against the confinement of their split-level homes while men commuted to and from the office with *Playboy* in their briefcases. Given these kinds of ambivalence, the baby-boom household failed to match the prescriptions of cultural opinion leaders. As with public opinion about communism, there was a discrepancy between the realms of the elite and the majority. The Cold War may have been a virus infecting the home front, but it did not create an all-consuming plague. However much the elite was occupied with mobilizing the nation against communism, ordinary women and men were dealing with private concerns that had at most an indirect relationship to the global crisis. "These are the tranquillized *Fifties,* and I am forty," wrote poet Robert Lowell.[42]

I think that the history of gender and family in the fifties was shaped primarily by the Depression and the World War. Thoughout the thirties, men and women had suffered economic hardship and insecurity, and had postponed marriage and children. With the war came jobs, marriages, and hopes for a happy future—but guarded hopes. Consumer goods were rationed, after all, and husbands were overseas. At last peace came, and with it an economic boom that allowed Americans to satisfy their deferred dreams. The baby boom and domesticity—and more generally, the craving for security—began before the Cold War and would have continued without it. Cold War policies nudged attitudes and behavior, but gender dynamics were shaped far less by national leaders than by what the public had experienced long before containment.

Films

A large number of movies in the 1950s dealt with troubled family relationships, but *My Son John* offered a distinctive explanation of what went wrong. "Breaking up homes," one character declares, is "a communist specialty." Whereas James Dean was a rebel with inept parents and without a cause, Robert Walker is a communist spy who lies to his parents (at one point with his hand on the Bible) about his efforts to subvert America. When his mother (Helen Hayes) finally uncovers the truth, she delivers a passionate monologue to John and an

FBI agent, hoping her son will confess. He doesn't, whereupon she collapses on a sofa, shouting "Take him away! He has to be punished!"[43]

My Son John belongs to what one is tempted to call a film rouge genre of perhaps fifty anticommunist propaganda films that came out of Hollywood during the postwar decade. The titles convey the fevered, heavy-handed approach: *The Iron Curtain* (1948); *The Red Menace* (1949); *The Steel Fist* (1952); *I Was a Communist for the FBI* (1951); *I Married a Communist* (1949); *I Led Three Lives* (1952).[44] Just as Hollywood studios had made their patriotic contribution during World War II, now they were doing it again in the Cold War. A particular incentive came from the 1947 hearings by the House Un-American Activities Committee seeking to uncover communists in the film industry. Producers hastened to prove their loyalty not only by purging suspect directors, actors, and writers, but also by putting ideological orthodoxy on the screen. (The director of *My Son John,* Leo McCarey, was a particularly cooperative witness before the Committee and later joined Cecil B. DeMille in demanding that members of the Screen Directors Guild sign a loyalty oath.) In these films the communists incite riots among Negroes and workers, beat up innocent people, steal mail, and drink to excess, while a few alert citizens try to awaken their fellow Americans to the peril.[45]

If we widen our lens beyond the obvious evidence, though, we see that these films were neither typical nor popular. Most of them—made with small budgets and nonstars—were financial disasters, ending up as second features on double bills. The studios produced them to appease the HUAC and other right-wing critics. But most moviegoers wanted diversion from the ominous realities that they saw in "March of Time" newsreels. If the producers wanted to attract audiences away from that new rival, the television screen, they would have to clothe Cold War values as entertainment.[46]

No one was more adept and successful at doing this than Walt Disney. In the postwar decade, Disney extended his movie-making genius beyond animated fantasies to live-action adventures that reverberated with tried-and-true Americanism. Work, religion, individualism, progress, and patriotism: these basic tropes ran through stories that took place in the distant past *(The Story of Robin Hood)* and the future *(Twenty Thousand Leagues under the Sea),* in Scotland *(Rob Roy, The Highland Rogue),* the American South *(The Great Locomotive Chase)* and the West *(Davy Crockett),* with heroes who are young *(Johnny Tremain)* or adult *(The Sign of Zorro)* or both *(The Swiss Family Robinson).* In every case, the protagonist—who is invariably male and of average socioeconomic status—confronts and overcomes an enemy to his society. "I think this is a good time to get acquainted with, or renew acquaintance with, the American breed of robust, cheerful, energetic, and representative folk heroes," Disney explained to a newspaper columnist in 1948. "They are worth looking at—soberly and in fun—to reeducate our minds and our children's minds to the lusty new world called America."[47]

Behind the robust and cheerful fun, some vehement political motives were goading Disney. He was still embittered by the strike in 1941 at Disney Studio. Technicians, support staff, and secondary artists had protested low wages, poor working conditions, and inadequate respect for their aesthetic contributions. Disney replied not only by hiring strikebreakers and fifty private police, but also by assailing the strike leaders as communists. So when the HUAC invited Disney to testify in 1947 about communist influence in Hollywood, he eagerly accepted. Although "at the present time everybody in my studio is 100 percent American," he said, communists had temporarily managed to "take over my artists" during the strike and continued "hiding behind this labor thing." After naming a few individuals and organizations, Disney exclaimed: "I feel that they really ought to be smoked out and shown up for what they are."[48]

Disney was an antiunionist, anticommunist businessman, but his business was entertainment. Except for these occasional outbursts to the press, he tried to "reeducate our minds" symbolically and smilingly: Uncle Walt to the child in each of us.[49] In *The Swiss Family Robinson,* a pious, hardworking family of father, mother, and three sons—emigrating to America—become stranded on an uninhabited island. During the next two hours we watch them building a microcosmic colony: crops; an elaborate treehouse with stove, running water, and elevator; domesticated animals (including an ostrich and elephant); a swimming hole; and a hilltop fortress. But all is not safe in paradise. The sons discover a band of pirates on the other side of the island. When the pirates attack, the Robinsons are prepared, defending themselves with a collapsible bridge, coconut bombs, ersatz land mines, and automated crossbows until finally they are saved by a warship.[50]

According to Disney's latest and best interpreter, Steven Watts, *Swiss Family Robinson* "neatly summarized the studio's Cold War vision." Watts says the contest between the virtuous, self-reliant citizens and the bloodthirsty, collectivist enemy—typical of so many Disney films—is a parable of Americans' postwar hopes and fears. A few columnists and congressmen at the time also discerned political meanings in the family films. "These pirates were oriental types, a collection of uglies with slant-eyes and bare bellies," wrote a Greensboro, North Carolina, columnist in 1961. "I only hope Western Civilization generally, when its time comes, can do half as well as the Swiss Family Robinson. Walt Disney, I suspect, could help Dean Rusk in some of his diplomatic problems with underdeveloped nations. We may need all of his ingenuity—and more—before it's finished."[51]

The movie reviewers, on the other hand, saw none of these deeper, darker meanings. "This grand adventure yarn," wrote the *New York Times* critic, is "a rousing, humorous, and gentle-hearted tale of family love amid primitive isolation and dangers." According to the *New York Herald Tribune,* "Walt Disney's holiday treat for the children [is] largely cheer and gadgetry with interludes of excitement." The *Saturday Review* found it "genial."[52] Given this innocent reading by movie critics (who were presumably opinion leaders), it seems farfetched to imagine that moviegoers—even those older than twelve—equated pirates with communists or coconut bombs with ICBMs. Just as Freud said that a cigar is sometimes only a cigar, a 1950s movie was most often only a movie.

Even though I resist labeling these popular entertainments as expressions of Cold War geopolitics, I certainly believe they were satisfying many Americans' psychological needs. Norman Vincent Peale's *Power of Positive Thinking* would not have been a best-seller for two years in a row if millions of readers hadn't been contending with some negatives. The rate of church attendance would not have reached an all-time high in 1957 if Americans weren't yearning for some larger reassurance. One of every three prescriptions would not have been for tranquilizers if the pill-takers, predominately women, were not troubled.[53] In this context, the resolutely optimistic formulas of popular entertainment certainly reflect the Zeitgeist. When they watched *Father Knows Best* or *Gunsmoke* or *Victory at Sea* during their daily five-and-a-half hours of television, or when they read *Marjorie Morningstar* or listened to Tony Bennett and Rosemary Clooney, they sought a happily-ever-after ending.

And when they visited Disneyland (as three million did in the first year it opened and five million each year thereafter), they could make-believe that their most innocent dreams would come true.[54] As a Florida wife and mother wrote to Disney in September 1961, shortly after the Soviets erected the Berlin Wall: "I must concede we are a confused people, with the clouds of war all around us. We're sick, and we're an anxious people who desire with all our might to leave a good world for our children and their children. Thank you so much, Mr. Disney, for working so diligently. . . . Your efforts have borne good fruit!"[55]

Generations

Finally we need to consider historians' favorite theme, change over time, and the enigma it raises. Regardless of whether they portray the fifties as nightmarish or confident or a blend of the two, scholars concur that most Americans were quiescent. So how do we explain the fact that a few years later—hardly a moment in cultural time—hundreds of thousands of young citizens were protesting not only Cold War policies in Cuba and Vietnam, but also the Cold War cultural stereotypes? Civil rights demonstrators went limp, antiwar men refused to fight, feminists challenged domesticity, and gays and lesbians came out of the closet. The Cold War was not over—in fact, was turning hot as napalm—and yet the security-state managers were losing control. How do we explain this surprising turnaround?

One answer is to trace the roots of the sixties' dissidence back into the apparently quiescent fifties and thereby minimize the change. Before 1960, we discover feminists and gay rights activists operating in a kind of underground. Lenny Bruce and the Beat writers were shouting their heresies in smoky cafes. African Americans were boycotting buses in Montgomery and taking their children to white schools in Little Rock. The more we search for ancestors of the 1960s' dissidents, the more of them we find, bleaching out the contrast between the two decades.[56]

A better answer is to trace back not to ancestors but to the formative experiences of the dissidents themselves. This is what psychologists call the life-course analysis, linking historical events to people's attitudes at successive phases of their lives.[57] It also allows historians to organize the past in terms of distinctive "generations," or, more precisely, cohorts.[58] Just as the group who came of age during the Depression and World War II espoused security and conformity in the postwar era, some of their sons and daughters coming of age during the early Cold War turned to dissent.[59] Seen through this social-psychological framework, cultural history takes place by delayed effect.

The theory of generations can lead us into the same kind of overgeneralization against which I've been writing this chapter. Just as there was more than one mood during the early Cold War years, there were several cohorts—young, middle-aged, old—each potentially divisible into narrower cohorts. What's more, each was differentiated by race, gender, and especially class. In his book about children on the World War Two home front, William Tuttle skillfully demonstrates how to juggle these many variables.[60] I can only sketch a hypothesis for future scholars to complicate and modify.

The spokespersons of Cold War culture, both the opinion leaders and the general public, were born in the 1920s and 1930s and were shaped by the Depression and the war. Contrary to what one might suppose, these two emergencies exerted positive effects. There was unemployment, disruption of personal lives, and the dread of being wounded or killed. Soon after the war ended, though, the survivors were enjoying college education (for men via the GI Bill) and bright futures. Given the "lessons" of their history, they dedicated themselves to security: on the one hand, fearful of losing it again; on the other hand, confident of its benefits. In turn, their children were shaped by and faithfully perpetuated this legacy of security. Among a certain segment of Cold War youth, the legacy produced unexpected outcomes.

The so-called generation gap of the 1960s was mostly mythical. Only on two issues did adults and youth split. A sexual revolution occurred in the later sixties, during which younger men and women enjoyed a freedom of attitude and behavior that alarmed if not appalled their parents. Meanwhile, the Vietnam War produced another generational disagreement, but not what one might think. Despite those vivid images of protesters placing flowers on National Guardsmen's bayonets or storming draft centers, people in their twenties were *more* supportive of the Vietnam War than were older Americans.[61] Only one of four campus students participated in a demonstration, and the majority of young people did not go to college.[62] In other words, if we focus on the majority of youth, we find a continuity of nondissidence between the fifties and sixties.

The dissident minority nevertheless exercised disproportionate influence on the course of events. Significantly, they too were an elite. They not only came from relatively affluent families and attended selective colleges and universities; they were campus political leaders. To borrow Halberstam's description of the Kennedy-Johnson administrations, they were the best and the brightest . . . and they defied the administrations' policies. Instead of Cold War containment, they advocated détente and opposed the war. Instead of top-down liberalism, they practiced grassroots democracy on behalf of blacks, the poor, students, and women. As the SDS Port Huron Statement proclaimed in the name of "the people of this generation": "We would replace power rooted in possession, privilege, or circumstance by power and uniqueness rooted in love, reflectiveness, reason, and creativity."[63]

We shouldn't misunderstand this intergenerational battle. Dissidents were carrying out, not contradicting, the precepts of the older cohort. Brought up by the half-fearful, half-confident lessons of "security," they adopted the confident half. Given material security (and Dr. Spock's respectful child-rearing), they grew up expecting to achieve their aspirations. "Self-fulfillment," they called it. And when they were thwarted, they responded with indignation. When the management refused to serve black college students a cup of coffee, they mobilized a civil rights movement. When the dean told Berkeley students to curtail their political activities, they launched a Free Speech Movement. When Johnson expanded the war, college men avoided or resisted the draft and their female allies cheered them on ("girls say yes to boys who say no"). And with an ironic twist, when men in these radical movements deprecated their female comrades, the women indignantly formed a movement of their own.

The feminist movement is not all there is to say about gender. Although left-wing men were in many ways as "tough" as the Cold Warrior ethic, they also exhibited soft, "feminine" qualities. Going limp, letting themselves be beaten and dragged by (long) hair to police wagons violated the John Wayne manly code. Refusing to fight for their country was even more unmanly. There were signs of androgyny not only on the left. Among male students who entered Harvard and Stanford in 1959 and 1960, an unusual number tended toward social work and psychology rather than math, science, and medicine. A notable proportion of this male cohort (born between 1941 and 1945) had a pattern of SAT scores that resembled women's: higher verbal than math. During the next two decades, gender differences on SAT scores shrank steadily.[64]

From this impressionistic evidence I derive a bold, somewhat speculative interpretation about a second Cold War generation. When they were growing up in the fifties, they were scorned as "goalkeepers," conformists, "the silent generation." Yet as many of them demonstrated (literally and figuratively) in the sixties, "security" impelled them not toward insecurity but toward risk. Reared in abundance, they felt empowered to claim their aspirations. Watching their mothers and fathers live out traditional gender roles with hints of ambivalence, the children felt safe enough to take the risks their parents shied away from. As they did so, the Cold War culture turned against itself.

Conclusion

So what should teachers tell their students about the Cold War at home, and what in turn should students tell their grandparents? Contrary to the usual claims by historians, Americans—at least nonelite Americans—were not obsessed by fear of a communist tidal wave overseas or a communist epidemic in their midst. On the other hand, those grandparents in retrospect, as well as the pollsters' respondents at the time, reported only part of the truth. They may have been more worried about grocery prices than the fate of Czechoslovakia or the treason of Alger Hiss. They may have enjoyed *The Swiss Family Robinson* as a good family yarn, not a dark parable of East vs. West. The Cold War nevertheless infiltrated their lives and their psyches. Its effects were diffuse, subtle, and subconscious. We can detect them in the surging use of tranquilizers, for example, and the crowded churches, and housewives' complaints about a problem they couldn't name.

The story of Cold War culture is more ambiguous and complicated than

Americans then or since have recognized. Let's continue exploring this fascinating part of our heritage.

Acknowledgments

I want to thank, first of all, Michael Hunt, who gave me generous and invaluable critiques. Many thanks also to Linda Orr, Dorrie Casey, Angela Davis-Gardner, Laurel Goldman, and Peggy Payne for helping this chapter go straight when it was going crooked. I'm also grateful to Peter Kuznick and James Gilbert for their support.

Notes

1. "Cold War culture" is a standard phrase.
2. Geoffrey S. Smith, "Commentary: Security, Gender, and the Historical Process," *Diplomatic History* 18 (Winter 1994), 90.
3. Stephen Whitfield, *The Culture of the Cold War* (Baltimore: Johns Hopkins University Press, 1991), 205. See also Tom Engelhardt, *The End of Victory Culture: Cold War America and the Disillusioning of a Generation* (Amherst, Mass.: University of Massachusetts Press, 1995).
4. Douglas T. Miller, *Visions of America: Second World War to the Present* (St. Paul: West Publishing, 1988), 54, 59, 72. I am not singling out this book for criticism. It happened to be the one I had assigned, but others say much the same.
5. David Caute, *The Great Fear: The Anti-Communist Purge under Truman and Eisenhower* (New York: Simon & Schuster, 1978); Richard Fried, *Nightmare in Red: The McCarthy Era in Perspective* (New York: Oxford University Press, 1990); Marty Jezer, *The Dark Ages: Life in the U.S., 1945–1960* (Boston: South End Press, 1982).
6. John Patrick Diggins, *The Proud Decades: America in War and Peace, 1945–1960* (New York: Norton, 1988); William L. O'Neill, *American High: The Years of Confidence, 1945–1960* (New York: Free Press, 1986).
7. On the contrary, people tend to remember public events that occurred when they were age 18 to 25: Howard Schuman and Jacqueline Scott, "Generations and Collective Memories," *American Sociological Review* 54 (June 1989), 359–81.
8. Peter Filene, "Recapturing the Thirties: History as Theater," *Change: The Magazine of Higher Learning* 6 (February 1974), 40–44.
9. According to postwar polls, southerners and rural residents were less attentive to foreign affairs: Gabriel A. Almond, *Americans and Foreign Policy* (New York: Frederick A. Praeger, 1960), 131–32.
10. V. O. Key, *Public Opinion and American Democracy* (New York: Alfred A. Knopf, 1964), 352–53, 371; Carl Solberg, *Riding High: America in the Cold War* (New York: Mason and Lipscomb, 1973), 43. By 1989, only half of Americans read a newspaper: Floris W. Wood, ed., *An American Profile: Opinions and Behavior, 1972–1989: Opinion Results . . .* (Detroit: Gale Research, 1990), 844.

11. Almond, *Americans and Foreign Policy,* 138–39; Michael Leigh, *Mobilizing Consent: Public Opinion and Foreign Policy, 1937–1947* (Westport, Conn.: Greenwood Press, 1976), 4–5.
12. Almond, *Americans and Foreign Policy,* 70–76.
13. Almond, *Americans and Foreign Policy,* Table 1, 73; *Gallup Poll: Public Opinion, 1935–1971,* 3 vols. (Wilmington, Del., 1978), 2:857, 1225, 1447.
14. Arthur Vandenberg, quoted by Robert Ferrell, *Harry S. Truman: A Life* (Columbia: University of Missouri Press, 1994), 251.
15. Quotation by John Lewis Gaddis, *The United States and the Origins of the Cold War, 1941–1947* (New York: Columbia University Press, 1972), 350–51. On subsequent opinion, Almond, *Americans and Foreign Policy,* 76–77. Gaddis claims that the administration had been responding to public opinion. But as Leigh notes, if that was the case, Truman wouldn't have felt the need for shock therapy: *Mobilizing Consent,* 23–25.
16. For the metaphor of virus, see Geoffrey Smith, "National Security and Personal Isolation: Sex, Gender, and Disease in the Cold-War United States," *International History Review* 14 (May 1992), 307–37.
17. Whitfield, *The Culture of the Cold War,* 10. The district judge is quoted in Whitfield, 33.
18. *Gallup Poll,* 2:1201.
19. *Gallup Poll,* 2:1225, 1240–41. When asked in 1954 specifically about the danger presented by American communists, 19 percent said "very great" and 24 percent "great," but 43 percent said only "some" or "hardly any": Samuel A. Stouffer, *Communism, Conformity and Civil Liberties: A Cross-Section of the Nation Speaks Its Mind* (Garden City, N.Y.: Doubleday, 1955), 59.
20. *Gallup Poll,* 2:1166–67.
21. *Gallup Poll,* 2:1203.
22. See chap. 5 by Meyerowitz in this volume.
23. Amy Kaplan, "Commentary: Domesticating Foreign Policy," *Diplomatic History* 18 (Winter 1994), 97–105. See also Emily Rosenberg, "'Foreign Affairs' after World War II: Connecting Sexual and International Politics," *Diplomatic History* 18, 59–70.
24. John F. Kennedy, "The Vigor We Need," *Sports Illustrated* 17 (July 16, 1962), 12; Johnson quoted in David Halberstam, *The Best and the Brightest* (New York: Random House, 1972), 532. See also Michael Kimmel, *Manhood in America: A Cultural History* (New York: Free Press, 1996), 267–70.
25. Donald J. Mrozek, "The Cult and Ritual of Toughness in Cold War America," in *Rituals and Ceremonies in Popular Culture,* ed. Ray B. Browne (Bowling Green: Bowling Green University Press, 1980), 178–91; Marc Fasteau, "Vietnam and the Cult of Toughness in Foreign Policy," in *The American Man,* ed. Elizabeth H. and Joseph H. Pleck (Englewood Cliffs, N.J.: Prentice-Hall, 1980), 377–415.
26. R. M. Marshall, "Toughening Our Soft Generation," *Saturday Evening Post* 235 (June 23, 1962), 13–17.

27. Arthur M. Schlesinger Jr., *The Vital Center* (Boston: Houghton Mifflin, 1963; orig. ed., 1948), 40–41.
28. Whitfield, *Culture of the Cold War,* 34–36.
29. Quoted by Whitfield, *Culture of the Cold War,* 43–44.
30. Fried, *Nightmare in Red,* 166–68; John D'Emilio, *Sexual Politics, Sexual Communities: The Making of a Homosexual Minority in the United States, 1940–1970* (Chicago: University of Chicago Press, 1983), espec. 24–31. "Lavender scare" comes from Smith, "National Security," *International History Review* 14 (May 1992), 307–37.
31. Elaine Tyler May, *Homeward Bound: American Families in the Cold War* (New York: Basic Books, 1988).
32. May, *Homeward Bound,* 103–8.
33. Talcott Parsons, "The Social Structure of the Family," in *The Family: Its Functions and Destiny,* ed. Ruth Nanda Ashen (New York: Harper Brothers, 1959), 271.
34. Quoted in May, *Homeward Bound,* 183.
35. Peter G. Filene, *Him/Her/Self: Gender Identities in Modern America* (3d ed., Baltimore: Johns Hopkins University Press, 1998), 181–83.
36. Edward A. Strecker, "Pops and Popism," *Parents' Magazine* 29 (February 1954), 119.
37. "The New American Domesticated Male," *Life* 36 (January 4, 1954), 42–45.
38. Michael Kimmel, *Manhood in America: A Cultural History* (New York: Free Press, 1996), 252–53.
39. Filene, *Him/Her/Self,* 177–80; Arthur M. Schlesinger Jr., "The Crisis of American Masculinity," in *The Politics of Hope* (Boston: Houghton Mifflin, 1963), 237–46.
40. "The Screen," *New York Times* (August 16, 1950). For other quotations and valuable analysis, see Steven Watts, *The Magic Kingdom: Walt Disney and the American Way of Life* (New York and Boston: Houghton Mifflin, 1997), 296–300.
41. Barbara Ehrenreich, *The Hearts of Men: American Dreams and the Flight from Commitment* (New York: Doubleday/Anchor, 1983), chap. 4; Alan Nadel, *Containment Culture: American Narrators, Post-Modernism, and the Atomic Age* (Durham, N.C.: Duke University Press, 1993), 129–36.
42. Robert Lowell, "Memories of West Street and Lepke," *Life Studies* (New York: Farrar, Straus and Cudahy, 1959), 85.
43. Quoted in Nora Sayre, *Running Time: Films of the Cold War* (New York: Dial Press, 1982), 94, 97.
44. Whitfield, *Culture of the Cold War,* 133.
45. Sayre, *Running Time,* 94.
46. Sayre, *Running Time,* 79–80; Whitfield, *Culture of the Cold War,* 133.
47. Quotation from Hedda Hopper, *Chicago Tribune* (May 9, 1948), quoted in Watts, *Magic Kingdom,* 289–90.
48. Quoted by Marc Eliot, *Walt Disney: Hollywood's Dark Prince* (New York: Carol Publishing Group, 1993), 191–94. On the strike and the HUAC, see Watts, *Magic Kingdom,* 98–99, 202–14, 225–26.
49. On his public statements, see Watts, *Magic Kingdom,* 346–49.

50. Watts, 301–2; Leonard Maltin, *The Disney Films* (New York: Crown Publishers, 1973), 177–79.
51. Watts, *Magic Kingdom,* 284, 300 and, more generally, chap. 15. W. D. S., "Indomitable Swiss," *Greensboro, N.C., Daily News* (January 3, 1961), quoted by Watts, *Magic Kingdom,* 302. On *Davy Crockett* as a "hero for the Cold War," see J. Fred MacDonald, *Television and the Red Menace: The Video Road to Vietnam* (New York: Praeger, 1985), 137–38.
52. Howard Thompson, *New York Times* (December 24, 1960), 8:4; Arthur Knight, "Make 'Em Laugh," *Saturday Review* 43 (December 3, 1960), 36. Likewise, "New Picture," *Time* 77 (January 13, 1961), 47. Maltin sums up the film as "hugely entertaining": *Disney Films,* 179. Likewise, *Time* described the early live-action films as "lightsome, modest," each being "a children's classic": "Father Goose," *Time* 64 (December 27, 1954), 46.
53. William E. Leuchtenburg, *A Troubled Feast: American Society since 1945* (rev. ed., Boston: Little, Brown, 1979), 104.
54. On television viewing, MacDonald, *Television and the Red Menace,* 112, 138–40, 147. On Disneyland attendance, Richard Schickel, *The Disney Version: The Life, Times, Art and Commerce of Walt Disney* (rev. ed., London: Pavilion Books, 1986), 316; and Watts, *Magic Kingdom,* 394.
55. Dorothy Mealy to Disney, September 23, 1961, quoted in Watts, *Magic Kingdom,* 350.
56. Joanne Meyerowitz, *Not June Cleaver: Women and Gender in Postwar America, 1945–1960* (Philadelphia: Temple University Press, 1994); Susan Lynn, *Progressive Women in Conservative Times: Racial Justice, Peace and Feminism, 1945 to the 1960s* (New Brunswick, N.J.: Rutgers University Press, 1993); Jezer, *Dark Ages,* chaps. 12–13; Taylor Branch, *Parting the Waters: America in the King Years, 1954–1963* (New York: Simon & Schuster, 1988).
57. Glen H. Elder Jr., "Social History and Life Experience," in *Present and Past in Middle Life,* ed. Dorothy H. Eichorn et al. (New York: Academic Press, 1981), 3–31; William M. Tuttle Jr., "America's Home Front Children in World War II," in *Children in Time and Place: Developmental and Historical Insights,* ed. Glen H. Elder Jr., John Modell, and Ross D. Parke (Cambridge, Mass.: Cambridge University Press, 1993), 31.
58. Norman B. Ryder, "The Cohort as a Concept in the Study of Social Change," *American Sociological Review* 30 (December 1965), 843–61; Alan B. Spitzer, "The Historical Problem of Generations," *American Historical Review* 78 (December 1973), 1353–85; and Karl Mannheim's classic chapter in *Essays on the Sociology of Knowledge,* ed. Paul Kecskemeti (New York: Oxford University Press, 1952), 176–322.
59. Glen H. Elder Jr., and Tamara K. Hareven, "Rising above Life's Disadvantage: From the Great Depression to War," in *Children in Time and Place,* ed. Elder et al., 47–72; Elder, *Children of the Great Depression: Social Change in Life Experience* (Chicago: University of Chicago Press, 1974).
60. William M. Tuttle Jr., *"Daddy's Gone to War": The Second World War in the Lives of American Children* (New York: Oxford University Press, 1993).

61. William L. Lunch and Peter W. Sperlich, "American Public Opinion and the War in Vietnam," *Western Political Quarterly* 32 (March 1979), 32–34.
62. *New York Times* (May 24, 1969), 68:3.
63. "The Port Huron Statement," in *The New Radicals: A Report with Documents,* ed. Paul Jacobs and Saul Landau (New York: Random House, 1966), 150, 153, 154.
64. These shifts were also correlated with father-absence during the war. For a detailed analysis, see Tuttle, *"Daddy's Gone to War,"* 225–30, and Tuttle, "America's Home Front Children," in *Children in Time and Place,* ed. Elder et al.

CHAPTER 8

Still the Best Catch There Is

Joseph Heller's *Catch-22*

Stephen J. Whitfield

year after his first novel appeared, Joseph Heller got a query from his Finnish translator, who needed to solve the following riddle: "Would you please explain me one thing: What means Catch-22? I didn't find it in any vocabulary."[1] By 1974 the translator could have consulted *Webster's New World Dictionary of the American Language* (2d College Edition), which classifies "catch-22" as a common noun. Yet only an uncommon author could coin so indispensable a term, and much about his book is unusual. It was the first novel Heller ever tried writing; and though the first chapter had been published in 1955 (in *New World Writing,* issue number 7), six more years were needed to finish the book. Had he known how long it would take, Heller later remarked, he might not have started writing it.[2]

The reviews were mixed. A few were even savage toward a work that seemed destined to suffer the oblivion so often inflicted on first novels. Later, for example, an unknown named Peter Benchley did not bother coming up with a title until minutes before his manuscript had to go to press. He could be cavalier because "nobody reads first novels anyway." (The book was *Jaws.*) Heller avoided that fate when, almost by chance, S. J. Perelman was being interviewed by the now-defunct *New York Herald Tribune.* Having once assured his daughter that "you can be as deeply moved by laughter as you can by misery," the humorist praised *Catch-22* as a terrific recent book. Yet even then it never came close to the *Times* best-seller list, living precariously as a word-of-mouth "cult" novel. It did rise to the top of the best-seller list in Britain, however, before catching on at home.[3]

As the military intervention in Vietnam gained momentum, and as that disaster helped to spawn a counterculture, the novel became a phenomenal popular success, guaranteeing that Heller would never need to dig for quarters out of car seats. A decade later *Catch-22* was more popular than immediately after publication, and it dwarfed the later, growing success of other serious novels that had appeared around the same time—like Ken Kesey's *One Flew Over the Cuckoo's Nest* (1962) and Kurt Vonnegut's *Cat's Cradle* (1963). In fact *Catch-22* is one of the most popular serious novels ever written, with well over twenty million copies sold in hardcover and paperback.[4]

The challenge of discussing *Catch-22* is daunting, since Canadian woodchoppers are still hacking down forests to facilitate critical evaluations of Heller's novel. The task undertaken here is different: it is not to try to illumine the text but instead to explore the context, to weigh the politics—not the aesthetics—of *Catch-22*, and to situate it historically as a way of understanding how its significance grew.[5]

Critics considering the structure and form of the novel have sometimes doubted whether such a designation is even deserved. Asked for a public comment, the most admired comic writer in English called *Catch-22* nothing but "a collection of sketches" that are "often repetitious—totally without structure," though Evelyn Waugh admitted that "much of the dialogue is funny." Rather than provide Simon & Schuster with a blurb, he advised the publisher to cut the manuscript "by about a half. In particular the activities of 'Milo' [Minderbinder] should be eliminated or greatly reduced."[6] Repetitive features that Heller deliberately injected also antagonized Norman Mailer, whose own first novel—also set in the war that he had directly experienced—conformed to the realistic standards for such fiction that *Catch-22* smashed. Yet so bloated in its circularity and haphazard in its arrangement did Heller's book seem that Mailer compared it to "yard goods," which "could [be] cut . . . anywhere. One could take a hundred pages from the middle of *Catch-22* and not even the author could be certain they were gone."[7]

Critics have also wondered whether Heller's book might be classified as a romance-parody, or as a mock-epic, and have noted its riffs on miscommunication—especially on how language can make the unbearable acceptable, and can insulate against the perception of injustice. But Heller's book also seemed too ribald, too stuffed with gags, to perpetuate the tradition of antiwar fiction. *Catch-22* was populated with "characters whose antics were far loonier than anything ever seen before in war fiction—or, for that matter, in *any* fiction." From Ernest

Hemingway and John Dos Passos down to Mailer, critic John Aldridge argued, war novels were supposed to be written in the vein of spare, austere "documentary realism."[8] But what was Heller getting at? Was he joking about the most horrifying of all themes, turning on laughing gas to get rid of the stench of death? The characters he had invented were mostly cartoons, and some were grotesque. The situations were outlandish. Lip-readers might well have inferred the following message: the men who defeated the Axis in the most awful and far-reaching of wars were uncomprehending buffoons whose commanding officers were either mad or moronic. Infantry platoons were often celebrated as rainbow coalitions, yet why would the protagonist of Heller's novel be an Assyrian and the chaplain an Anabaptist? No wonder then that Aldridge claimed that critics and other readers had to learn to become more sophisticated, to fathom the striking originality of Heller's novel.

Yet many of Heller's fellow writers were quick to understand and to welcome what he had accomplished. *From Here to Eternity* (1951) barely resembled *Catch-22.* Yet James Jones praised it as "a delightful and disturbing book. Its weird comedy is marvelous, and underneath this on an entirely different level, its pathos for the tragic situation of the men is equally fine." Two years after it was published, Dos Passos himself, then a conservative and a Republican, also spoke highly of so subversive a text. The author of *The Young Lions* (1948), Irwin Shaw, also raved about Heller's new novel. The oeuvre of John Steinbeck included *The Moon Is Down* (1942), set in World War II; and no American novelist had done more for social realism. Yet the Nobel laureate realized that *Catch-22* merited rereading, that "a good book" the first time around proved to be "loaded with things that must be come at slowly. . . . My wife says she knows when I am reading *Catch-22* because she can hear me laughing in the next room and it is a different kind of laugh." Old-fashioned sorts of writers were quick to pick up the radically disorienting portrayal that this first novel was presenting, but this litany would be incomplete if it did not mention the impact that *Catch-22* exerted on a 23-year-old unknown whom Heller's agent claimed was "the only other genius" she had the privilege of representing. From the future author of another novel set in World War II, *Gravity's Rainbow* (1973), Candida Donadio received a letter raving about Heller's book. Thomas Pynchon declared it among the best novels he had ever read.[9]

In the early 1960s Heller also received unsolicited fan mail from Nelson Algren ("the laughter is hard-won"), from Jeremy Larner ("I read every word of *Catch-22* with great delight and ended up scared and moved and happy"), from the British drama critic Kenneth Tynan ("a bloody masterpiece"), and from Stephen Ambrose, who would become a prolific military historian: "For sixteen years I have been waiting for the great antiwar book which I knew World War II must produce. I rather doubted, however, that it would come out of America; I would have guessed Germany. I am happy to have been wrong." In some ways, Ambrose added, *Catch-22* is superior even to *All Quiet on the Western Front* (1929).[10] The enthusiastic appreciation of other writers Heller earned at the outset.

But as Aldridge also acknowledged, the enhanced status of *Catch-22* was not only a matter of professional recertification. Political events caught up with this novel, and would tutor its readers in how chaotically the world is structured: "It was undoubtedly this recognition that the book was something far broader in scope than a mere indictment of war—a recognition perhaps arrived at only subconsciously by most readers in 1961—that gave it such pertinence to readers who discovered it over the next decade. For with the seemingly eternal and mindless escalation of the war in Vietnam, history had at last caught up with the book and caused it to be more and more widely recognized as a deadly accurate metaphorical portrait of the nightmarish conditions in which the country appeared to be engulfed."[11] It became increasingly apparent that *Catch-22* answered with gusto the famous complaint that Philip Roth had uttered in that same year, 1961: "The American writer in the middle of the twentieth century has his hands full in trying to understand, and then describe, and then make credible much of the American reality. It stupefies, it sickens, it infuriates, and finally it is even a kind of embarrassment to one's meager imagination. The actuality is continually outdoing our talents."[12] An increasingly unintelligible context thus began to make a text intelligible.

An appreciation of Heller's achievement requires situating his novel in four distinguishable time periods: the experience upon which the author drew for much of the material, the period in which it was composed, the moment in which it was published, and the era that it so eerily anticipated—the era in which our own sensibility has inevitably been shaped.

He emphatically denied that *Catch-22* was about the Second World War. The B-25 bombardier who would create John Yossarian made no effort to avoid combat. Having enlisted in the Army Air Corps, Heller "actually *hoped* [to] get into combat. I was just nineteen and there were a great many movies made about the war; it all seemed so dramatic and heroic." Speaking puckishly at Brandeis University while the Vietnam War was raging in the fall of 1969, he told the doubtless puzzled undergraduates (as I recall): "I *liked* World War II." The B-25's, Heller recalled, "were ridiculously small. We were assigned to a squadron on the west coast of Corsica," with the 340th Bombardment Group of the 12th Air Force, and with bridges usually as the targets. "In ten months I flew sixty missions, and my squadron lost two or three planes at the most." Fear did not grip him until the thirty-seventh mission. "Until then it was all play, all games, it was being in a Hollywood movie. I was too stupid to be afraid." But a run over Avignon was dangerous; the gunner was severely wounded and bled heavily into his flight suit. Until then "war was like a movie to me." But he quickly realized, "Good God! They're trying to kill me too." Afterwards even the milk runs were scary. "Until Avignon, the war was the most marvelous experience in my life," he insisted. "It was wonderful . . . [But] war wasn't much fun after that." Petrified of flight, Heller came home as a first lieutenant and did not get airborne again for fifteen years.[13]

His novel was not intended to be "realistic," to make vivid for readers the ordeal of combat, and was designed neither to represent the actualities of World War II nor to downgrade its value. The theme of *Catch-22* is not arms, and the man who wrote it denied in 1968 any special "interest in the subject of war . . . I was interested in personal relationships to bureaucratic authority." (But was it accidental that the world's largest building was the Pentagon?) His own "attitude is against war,"[14] he acknowledged; yet World War II had taught him how easy it is to worship Mars. Admittedly "it *takes* a certain amount of courage to go to war, but not very much, not as much as to refuse to go to war."[15] The book that had inspired him as a child to want to become a writer himself was "a children's prose version of *The Iliad*," which he wanted to write himself when he grew up. But since Homer had already done it (and in verse to boot), Heller recalled being stuck with devising "a war novel of my own, containing heroes of imperfect character, adversaries with sympathetic, sometimes even likable faults, and a good many old people mourning the deaths of the young. Such thematic similarities may be coincidental; but the reference late in the book to Yossarian's sulking in his tent like Achilles and refusing to fight is not."[16] (Heller inflates the similarities. Very few of the faults his own characters exhibit seem pleasant, and deaths are rarely mourned in *Catch-22;* but never mind.)

Nor does the *Wehrmacht* get more than a glancing mention, because what affected its author was the political atmosphere of the Korean War and the Cold War. The intrusions of political orthodoxy and of bloated bureaucracies had injected absurdities into American society that contaminated the ideals of a moral life, that seemed to be corrupting everything. Organization men were generating chaos, incoherence, cruelty—not on Corsica (called Pianosa in the novel) but in his native land. Service in the Army Air Corps a decade earlier had instigated Heller's impulse to lampoon the structures of hierarchy and authority, and thus national habits of antinomianism were intensified and extended.

Catch-22 was even denounced by one reviewer as "immoral" for "being inclusively, almost absentmindedly, anti-institutional."[17] What was worthy of resistance was not the actual war Heller had fought in but the undeclared war that was stifling liberal impulses and making common experience tawdry, that was commanding unthinking assent: "The cold war is what I was truly talking about, not the World War." A novel that is punctuated with the Glorious Loyalty Oath Crusade, that has messages hidden in plum tomatoes rather than pumpkins, that has congressmen asking "Who promoted Major Major?" derives from the Cold War and its "frightening" impact "on the domestic political climate" in which an eponymous demagogue invariably carried a briefcase bulging with portentous files. "And that's the spirit of revolt that went into *Catch-22*."[18]

In interviews Heller made this point so often that his testimony deserves to be taken as reliable. Having taken notes as early as 1953, he began to write it "during the Korean War and aimed it for the one after that,"[19] as though fearing that war might be back by popular demand.

In 1961 the thirty-eight-year-old author was enjoying a successful career in magazine advertising, and his political attitudes and civilian experiences had left behind the eager teenager who had enlisted in the glamorous fight against fascism. The 1950s had stimulated him to tap impulses of opposition, to try to subvert convention by conceiving of a contrarian like Yossarian. What Heller had come to see as imperative was resistance to what the sociologist C. Wright Mills called in *The Causes of World War III* (1958) "crackpot realism," the lunacies of the bureaucratic mind that led the FBI, for example, to infiltrate the severely crippled Communist Party rather than to monitor organized crime (whose existence the bureau denied). The party was so honeycombed with influential informants that, by 1956, Communist policies could have been manipulated if not determined from FBI headquarters. Among the contributors to *The New Yorker* who merited FBI files were E. B. White, Edmund Wilson, and even Perelman, though the most subversive was Dorothy Parker, who provoked about a thousand pages of tax-supported surveillance.[20]

Such obsessions with the insignificant encouraged popular disenchantment and irreverence. Yet challenging the follies of power seemed pointless. By then patriotism had become so pervasive and complacent (even smug) that a wary defensiveness was necessary even for those who wished to register rational, democratic dissent. In 1960 Mills's colleague at Columbia University, Daniel Bell, felt obliged to insist upon what any sensible reading of American history might disclose, which is that "one can be a critic of one's country without being an enemy of its promise."[21] Political energies seemed so exhausted that the will to move even in liberal directions seemed paralyzed. The opportunity to mobilize constituencies for change seemed so restricted that *Catch-22* could hint at no social reforms and offer no prescriptions.

Its tone implied that one authentic response was laughter. In writing drafts of the novel, Heller was aware of humorous situations. But not until he heard a friend laughing out loud did Heller realize that "I could be comic," he recalled. "I began using that ability consciously—not to turn *Catch-22* into a comic work, but for contrast, for ironic effect."[22] He was as surprised that grisly themes could come out funny as was Stanley Kubrick, who had attempted around the same time to direct a serious film about the causes of World War III, only to realize that the geopolitical situation was so preposterous that *Dr. Strangelove* (1964) defied Bob Hope's dictum for comedy: "Certain things you can't touch. The atom bomb—you can't play with the atom bomb."[23] Though Heller "honestly didn't know I had the power to write something that would make people laugh," he denied bucking for the status of jester. He was not competing with Art Buchwald or Russell Baker, both of whom raved in private correspondence about *Catch-22,* because its facetiousness "was not the end but the means to an end."[24]

The term "catch-22" entered the language on comic wings, as though replacing the gloomier Central European adjective "Kafkaesque" as shorthand for bureaucratic madness. But the differences can be exaggerated, since Kafka himself broke out into laughter when reading his mordant tales aloud. If there is any exit from the terrifying indifference of a universe ruled by an absentee God, laughter may provide a map. In *Catch-22,* "the descent into phantasmagoric horror that occurs in the concluding chapters," Aldridge wrote, "is not a violation of the comic mode but a plausible vindication of it," which echoes what the critic Robert Brustein noticed as early as 1961: "The escape route of laughter [is] the only recourse from a malignant world." The author of *Catch-22* proved himself "a first-rate humorist who cripples his own jokes intentionally," Vonnegut remarked, "with the unhappiness of the characters who perceive them."[25]

The phenomenon of "black humor" was hardly invented in 1961; but it has served the era admirably, when war would spell disaster for our species and satisfy no rational political aims. The arms race has become so permanent a feature of the modern predicament that "peace on earth," as Heller would write in his fifth novel, "would mean the end of civilization as we know it."[26] Effortlessly tossing off such epigrams suggests one model: the joker is Wilde, who could also invert a cliché in a way that is wondrously perceptive rather than merely clever. Such a gift cannot be learned. But treating terrifying experiences flippantly, plucking mirth from the macabre—that is a flair that Heller claimed to have picked up not from Wilde but from Waugh, Céline, Nabokov, and Nathanael West. However identifiable the literary voices that echo in *Catch-22,* which Murray Kempton called "a staggeringly funny book relieved by horror,"[27] black humor—slapstick with a shudder—achieved an emancipating effect.

Though the 1960s is so often characterized in terms of youthful dissidence, campus radicals themselves were hardly created ex nihilo. They were influenced by figures older than "the kids" themselves, like Paul Goodman and Malcolm X (*né* Little), each of whom avoided service in World War II by making themselves into "stinker" cases whom military recruiters could not handle. A record of resistance to the Second World War, however, was not a requirement for affecting the young who spearheaded opposition to the war in Indochina.

Consider another Brooklynite like Mailer, who had been a rifleman in a platoon in the Pacific; or Dr. Herbert Marcuse, an analyst for the OSS (the predecessor of the CIA). Howard Zinn's *Vietnam: The Logic of Withdrawal* (1967) was the first book to urge immediate and unilateral departure from Indochina. In 1943, at the age of twenty, he had volunteered for the Army Air Corps, and had served as a bombardier with the 490th Bomb Group, which helped destroy the town of Royan (near Bordeaux) in mid-April. Probably more civilians were killed in that town than at Coventry or at Rotterdam, though few of the victims in Royan were foes. To inflict this atrocity, a new substance was used: "jellied gasoline," later called napalm. At the time the future Boston University professor neither voiced nor harbored any moral qualms about such bombardments.[28] But so dramatic a shift had occurred by the late 1960s, so irreverent had political attitudes become, that *Catch-22* could encapsulate the repudiation of regimentation that the military exemplified.

Had the novel come out much earlier than 1961, had it been published when its first chapter had appeared, so disorienting a stance might have buried *Catch-22* or confined it to a cult whose readers would have praised the novel as "prophetic." Even in 1961, its better-Yossarian-than-Rotarian impudence made it prematurely radical. Paul Goodman's publisher, for instance, had rejected *Growing Up Absurd* before Random House agreed to publish it in 1960;[29] its title conveyed a sensibility that would help make Heller's own novel canonical. The spitballs aimed at authority needed some time to hit their targets, and in 1961 the definition of those targets was beginning to shift.

One way of locating the milieu is to contrast two famous experiments in clinical psychology. In 1951 Solomon E. Asch announced his study of conformity, in which the individual submits to the superior authority of peers in perceiving the length of sticks. In 1960–1963, his student Stanley Milgram conducted experiments in which the individual yields to the superior authority of authority—which may be malevolent and sadistic. In the 1951 experiments, the naive subject accedes to the claims of the group, even to the point of denying "the unmistakable evidence" of his own eyes. In the Milgram experiments, the naive subject "comes to view himself as the instrument for carrying out another person's wishes, and he therefore no longer regards himself as responsible for his actions."[30]

The 1950s posed the problem of the tyranny of the majority; group pressures activated the anxiety that someone would be left with the short end of the stick. The 1960s presented another kind of tyranny, as policies were pursued outside of democratic restraint, as self-government seemed to transform itself into an irrationality that expected unthinking allegiance. Civic health thus meant sticking it to authority.

The historic shift that Heller's novel adumbrated seemed to discredit the prudence marking the decade when *Catch-22* was written. In 1962, the novelist privately expressed some surprise that no congressional committee subpoenaed him, nor did any officials wish him ill for the "rebellious and iconoclastic spirit" that *Catch-22* sanctions: "I don't think this is because I have been clever, but because there may be a much wider current of sophisticated and discontented people than many of us believe."

Wariness about criticizing the status quo was replaced by weariness with its inertia. Even utopianism bounced back to life—not only in Goodman's creative anarchism but even at pillars of the educational establishment, like Harvard. Between 1948, when Professor B. F. Skinner's novel, *Walden Two,* was published, and 1960, its sales totaled only nine thousand. In 1955, for example, only 250 copies had been sold. But in 1961, his tale of a visit World War II veterans make to an alternative community sold eight thousand copies; and by the mid-1960s, fifty thousand copies of *Walden Two* were being sold annually.[31]

Yet utopianism could somehow coexist with cynicism, optimism with disenchantment. Visiting her alma mater in the spring of 1971, Diana Trilling happened to ask a Radcliffe student about the Spanish Civil War. The reply illustrated a political tectonic shift: "I don't know much, but I do know enough to know it was our fault!" Here an amnesia that teachers of history have grown hoarse lamenting blended with a knee-jerk antipatriotism that simply inverted the star-spangled ideology of the previous decade, but the transformation was remarkable. In 1952 the *Saturday Evening Post* had excerpted the autobiography of Whittaker Chambers; little more than a decade later, the same magazine was serializing *The Autobiography of Malcolm X.* After examining six high school history textbooks, academicians Nathan Glazer and Reed Ueda concluded in 1983 that "moralism and nationalism are both out of date." Patriotism, which schooling was once supposed to inculcate, had evaporated.[32] With the rise of what Mrs. Trilling's husband called "the adversary culture," no consensus could be articulated with assurance.

Consider the fate of Paramount's version of *Catch-22.* It was released in 1970, the same year as *Patton* and *Tora! Tora! Tora!* (both set in the Second World War) and *M*A*S*H* (set in the Korean War)—which meant that the mass audience had fragmented. Its tastes could no longer be easily gauged. For the next six years not even antiwar films were made, with a few exceptions, because ideological fissures entailed marketing risks. Only in 1976 did Hollywood again begin to release films about the military. By then the war in Indochina was over, however slowly the scars were healing. Even though the cinematic *Catch-22* (like *M*A*S*H*) uses war only as a backdrop,[33] Mike Nichols's $15 million film bombed at the box office. Nor was it a *succès d'estime.* If the film "had been foreign, in black and white, without stars and based on an unknown novel," Heller conjectured, "it would have been a major critical success. This is not a comment on the quality of the film but on the consistency of film reviews." Having sold the screen rights, he had no personal involvement in the movie, which he nevertheless felt turned out "okay." But his own expectations were low. The novel that had already attracted five million readers was too sprawling and intricate to be transferred fairly intact to the screen; and Heller realized that much of his humor had been forfeited "in a vain attempt to establish a 'story line'—something that the novel didn't have to begin with."[34]

But another problem was Yossarian (Alan Arkin). "For a majority of the American people in 1970," Lawrence H. Suid speculated, "heroes didn't run away . . . even to live." They were supposed to fight and if need be die for their country; and paddling in the general direction of neutralist, socialist Sweden was not an option. The bombardier's disclaimer—that he was not escaping from his responsibilities but accepting the duty of saving himself—was for the mass audience a discordant and anachronistic message, since *Catch-22* certainly looked like it was set in the Second World War; and one of the world's largest fleets of bombers was assembled for the Paramount production. Diagnosed as suffering from "a morbid aversion to dying," Yossarian looked out of place at the center of a cinematic crusade in Europe. Film scholar Thomas Doherty has yet to see a "successful big-budget 'deconstruction' of the Second World War mythos on screen, and as long as historical memory can conjure the Army Signal Corps footage of the Holocaust there won't be."[35]

Readers may have been more receptive than moviegoers to the redefinition of heroism that *Catch-22* proposes; and disappointing box-office receipts should not obscure the cultural transformation that defines the 1960s. In his prescient review of "this explosive, bitter, subversive, brilliant book," Brustein did not flinch from the protagonist's preference for desertion over either flying more missions or allowing himself to be co-opted by Colonel Cathcart. Yossarian's decision, Brustein argued, was not only structurally sound—a fulfillment of the premises that the novelist had laid—but also a laudable gesture of "inverted heroism," which the critic called "one of those sublime expressions of anarchic individualism without which all natural ideals are pretty hollow anyway." Unlike the naive subjects whom Asch and Milgram studied, unlike nearly every other character in *Catch-22,* Yossarian accepts the notion of personal accountability even within a morally debased institution.[36]

His options are terribly constricted. "Accept the corruption and benefit by it, join us," Heller later explained, and "we'll send you home a hero; or else, go to prison for refusing to fly more missions; or fly more missions until you're eventually killed." Yossarian can only "assert himself, without accepting any of these obnoxious alternatives," if he can Just Say No. "Now he knows he's not going to get to Sweden. The novel ends with him going out the door."[37] That haven is remote (though hundreds of Vietnam war deserters did find refuge in Sweden). Yossarian chooses flight, like Huckleberry Finn. But no equivalent of Huck's adventures in the Indian territory awaits Heller's fugitive even if he does beat the odds. "Even if they don't find you," Major Danby warns, "what kind of way is that to live? You'll always be alone. No one will ever be on your side, and you'll always live in danger of betrayal." Yossarian's rebuttal is unanswerable, however: "I live that way now."[38]

What enables Yossarian to detach himself from the group? He is of course a loner, without parents or siblings or spouse or children, bereft of any sort of background that the novelist might have thickened. No personal loyalties might conflict with what Yossarian ultimately conceives as the primacy of his own survival; without a family to complicate his allegiances, his moral awareness is largely intact. Since Heller ascribed the idea of putting an outsider at the center from the pursuit of Leopold Bloom through Dublin, and since two of Heller's later novels—*Good as Gold* (1979) and *God Knows* (1984)—are "ethnic" in theme and idiom, it is tempting to wonder whether Yossarian is related by blood to Bruce Gold and King David. One preliminary sketch does make Yossarian Jewish. But his creator then "made him Assyrian (but what I was ignorant of, for one thing, [is that] his name is not Assyrian; I've since been told it's Armenian)." Heller "wanted to get an extinct culture, somebody who could not be identified either geographically, or culturally, or sociologically—somebody as a person who has a capability of ultimately divorcing himself completely from all emotional and psychological ties." Yossarian is an isolato. (Maybe Philip Roth's *The Great American Novel* was aiming at the same eerie status in making the rookie pitching sensation named Gil Gamesh a Babylonian.) Ethnicity is a sign that Yossarian doesn't quite belong, and perhaps fortifies his refusal to live as a submissive conformist. "The country was in peril," Minderbinder (a family man) warns Yossarian, who "was jeopardizing his traditional rights of freedom and independence by daring to exercise them." Such contrariness may elucidate Heller's remark that "Yossarian was very Jewish, but I didn't know that until years later."[39]

By making his protagonist a deserter who also deserves a medal, Heller helped contest the category of heroism. Admittedly its definition would be altered under the impact of the Vietnam War, which led even the highly decorated Audie Murphy to object in 1970 to "ask[ing] young men to risk their lives in wars they can't win. . . . What if my sons try to live up to my image? . . . I don't want them to be what I was. I don't want dead heroes for sons."[40] Lifetime civilian John Wayne had been exempted from Murphy's war. But General Douglas MacArthur told the star of *Sands of Iwo Jima* (1949), "You represent the American serviceman better than the American serviceman himself"; and Congress concurred by authorizing a John Wayne medal for having personified the nation's martial virtues.[41] Bravery under fire had ceased to be related to heroism. When a 1985 biography of another actor was subtitled "a hero's story," the term edged near absurdity. The subject of the biography, Sylvester Stallone, had avoided military service during the Vietnam War by teaching at a ritzy school in Switzerland and then, at the University of Miami, by studying the craft of acting (though the results have been disappointing). In 1995 the media were hailing a serviceman shot down in Bosnia as a hero—a status Captain Scott O'Grady had not earned by rescuing anyone, or by halting enemies in their tracks, but merely by surviving for six days. In the year *Catch-22* was published, *Life* magazine pushed beyond our own species in extolling "Ham"—the nation's first "astrochimp"—as a "real hero."[42] But surely the revisionism of novels like Heller's made it tougher to agree on who a hero is—or for that matter a "heroine." That sobriquet was bestowed by Patricia Ireland, the president of the National Organization for Women, upon Lorena Bobbitt, a victim of domestic violence who told the arresting officer of dismembering John Wayne Bobbitt because "he always have orgasm and he doesn't wait for me to have orgasm. I don't think it's fair so I pulled back the sheets and I did it."[43]

Revising or at least enlarging the notion of heroism was only one way that *Catch-22* anticipated the trauma in Vietnam. The novel also exposed the inequities of power and status that the ideology of a homogeneous American Way of Life disguised. Such inequities had been so subtly concealed that, in 1961, Richard H. Rovere's portrait of "The American Establishment" was greeted with some disbelief by readers of the *American Scholar*. The journalist's delineation of a permanent and covert *nomenklatura* was taken to be a satire, or a put-on. "Many readers professed to be puzzled. Some even asked if I intended my work to be taken seriously," he recalled. Admittedly much of it was tongue-in-cheek. Readers were informed that Communist Party boss Gus Hall and the Dodgers' Duke Snider, for example, were not members of the Establishment.[44] But Rovere's effort to record the composition and conduct of so undemocratic an elite represented a shock to the system.

In highlighting the sting of class differences, *Catch-22* could also claim fictional antecedents; readers could hardly forget Thomas Sutpen directed to the back door in *Absalom, Absalom!,* or Jay Gatsby dismissed as "Mr. Nobody from Nowhere."[45] Social inequality (which is separable from the injustices associated with race and gender) is a chasm which Faulkner's protagonist learns to measure near the beginning of his life, Fitzgerald's protagonist only at the end. But Heller's novel gives such knowledge a political kick: Colonel Cathcart is astonished that enlisted men worship the same God as officers—some of whom, it should be noted, are okay. Yossarian is after all a captain, and neither Major Major nor Dunbar nor Danby could be described as villainous. Their commanding officers—Generals Dreedle and Peckem and Colonels Cathcart and Korn, plus Captain Black—are dangerous dunces, however,[46] which is why *Catch-22* can be read as a swipe at discrepancies linked to the national creed of egalitarianism.

Published when the $140 billion intervention in Indochina was barely a rumor of war, Heller's novel made more sense of the social disparities of the 1960s and early 1970s than of the 1940s. The Second World War was a popular war which also approximated a people's war. To defeat the Axis, about ten million conscripts were required. But the privileged also volunteered—like George Bush, who became a Navy fighter pilot, and like three sons of former Ambassador Joseph P. Kennedy. The Boston Brahmin McGeorge Bundy quit Harvard's elite Society of Fellows to enlist; rejected for poor vision, he got in by memorizing the eye charts.[47] After the Japanese attack on Pearl Harbor, U. S. Senator Paul Douglas, a former professor of economics and a liberal Democrat, enlisted in the Marines, even though he was then fifty years old.

During the Vietnam War no senators or congressmen emulated him, nor did many of their sons. By 1970, 234 sons of senators and congressmen happened to be old enough to have served in the military. Yet according to journalist Myra MacPherson, "more than half—118—received deferments. Only 28 of that 234 were in Vietnam. Of that group, only 19 'saw combat.' Only one, Maryland Congressman Clarence Long's son, was wounded." No member of "the House Armed Services Committee had a son or grandson who did duty in Vietnam. Student deferments [and flunked physicals] were shared by sons and grandsons of hawks [like Barry Goldwater] and doves [like Alan Cranston] alike." In the mid-1960s, college graduates made up only 2 percent of Army inductees. In Harvard's class of 1970, according to one poll, were 1200 men, exactly two of whom served in Vietnam (fewer than the number of Kennedys who had served in World War II). Even in West Point's class of 1966, only one in six volunteered to serve in Vietnam.[48] Though Julie Nixon Eisenhower expressed her willingness during the 1972 campaign "to die for the Thieu regime,"[49] few of the privileged shared such brave sentiments.

Patriotic appeals to national unity and selfless service could not erase the stubborn fact of class difference. Though Ronald Reagan made a habit of praising the nobility of fighting communism in Vietnam, his own eldest son was exempted on medical grounds. Hawks like future Senator Phil Gramm (Republican of Texas) and Representative Newt Gingrich (Republican of Georgia) took advantage of student deferments, which did not prevent the then-future Speaker of the House from telling a reporter that Vietnam had been "the right battlefield at the right time." But when asked why it wasn't right for him, Gingrich got testy: "What difference would I have made? There was a bigger battle in Congress than in Vietnam"—though he was not elected to Congress until four years after the last grunts had left Saigon. A bum knee kept the hawkish Pat Buchanan out of uniform during the Vietnam War. A bad back kept the hawkish Elliott Abrams, the assistant secretary of state in charge of the Nicaraguan contra war under Reagan, a civilian in the 1960s. A married student's deferment prevented the bellicose pundit George Will from seeing any action either. Divinity school attracted David Stockman, whom Reagan later appointed director of the budget. Chopping the funding for Vietnam Veterans' Centers, Stockman minimized the claims of veterans, whom he called "a noisy interest group."[50] Nor did later conservative icons like Justice Clarence Thomas and Rush Limbaugh serve in Vietnam either.

For those who lacked Stockman's inclination toward theology, the National Guard beckoned and managed to attract more of the college-educated than all other services combined. The National Guard was the most racially homogeneous of the services; only 1 percent of its personnel was black. Yet among professional athletes, the National Guard appears to have been the service of choice—again, in striking contrast to the Second World War, when thirty-year-old Hank Greenberg enlisted in the air force less than a week after Pearl Harbor. The first baseman served until the summer of 1945, coming out a captain of a B-29 bomber squadron. At his peak Ted Williams disrupted his career with the Boston Red Sox by serving four and a half years as a fighter pilot, first for the Marines in the Second World War, and then again—though he was thirty-three and married and a father—flying jets in Korea (where a wingmate was John Glenn).

The chance to fight in Vietnam attracted no prominent athlete, however, though a billionaire's grandson who had once considered a career in professional golf would, in retrospect, prove to be the most famous member of the National Guard. Later, as a senator and a vice president, J. Danforth Quayle championed a strong defense. But by enlisting in the National Guard, Quayle forfeited his only chance to take up arms for his country. During the 1988 campaign, he justified his hell-no-we-won't-go decision by recalling an eagerness to go to law school, a momentary desire—since he eschewed that profession anyway. Also claiming to have disliked the "no-win" policy that the Johnson administration had been pursuing, Quayle nevertheless did not enlist in the National Guard until Nixon took office. It is permissible to speculate that Quayle's quandary would have disappeared had he been allowed to pay $300 to let a substitute serve (which is how both Grover Cleveland and the father of Theodore Roosevelt, for instance, avoided fighting for the Union during the Civil War). What is undeniable is Senator Quayle's voting record; he opposed increased benefits for veterans.[51]

Even for those unlucky enough to be assigned to duty in Vietnam, caprice was involved in getting the prized assignments to rear echelon units. In 1968 only 12 percent of the troops were assigned to line units—again unlike the Second World War, during which sacrifice was more widely shared, at 39 percent. Those who mostly fought in Indochina were the underprivileged: the working class and the lower class. They tended to have already been left behind in the competition for the right schools or for any schools at all; they had already lost out in the competition for good jobs and, if they survived the war, would come home to declining real wages. It is not surprising that some soldiers were dressed to kill by wearing helmets on which the phrase "catch-22" was painted. Others were seen carrying copies of the novel, or could identify with Yossarian, who realizes that "only a fraction of his countrymen would give up their lives [in this war] . . . and it was not his ambition to be among them. . . . History did not demand Yossarian's premature demise; justice could be satisfied without it." He concludes: "That men would die was a matter of necessity; *which* men would die, though, was a matter of circumstance, and Yossarian was willing to be the victim of anything but circumstance. But that was war."[52]

Heller doubted whether his novel affected anybody's military conduct in Vietnam: "It just confirmed their opinion that 'this is crazy! I don't know why we're here. And we'd better watch our superior officers because they can be as dangerous to us as the people out there.' That turned out to be true." *Catch-22* anticipated the redefinition not only of what a hero is, but also of who an enemy is—and it could mean anyone likely to get a soldier killed. In Vietnam assassinations or assassination attempts against unpopular officers or noncommissioned officers were known as "fraggings," because fragmentation grenades did not leave fingerprints. In 1969, 126 incidents were reported; two years later, the number of reported "fraggings" had nearly tripled, even as troops were being withdrawn.

"Vietnam was, for American soldiers, the perfect training ground for paranoia," according to historian Christian G. Appy. "To assume that everyone was a potential enemy was, in fact, a reasonable psychological response to the realities of counter-guerrilla warfare," which also "encouraged soldiers to suspect everyone, even other Americans," such as "officers bucking for promotion and all too eager to put their units in jeopardy to build up a good body count."[53] Vietnam was probably the first war in which American combatants did not generally fight to take an enemy's terrain or capital, probably did not fight for a cause or even for victory. Instead they imitated Yossarian. They simply wanted to survive the 365 days before rotation home, to get out of their tour of duty alive and in one piece. But unlike Yossarian, who knows that an enemy surrender is inevitable, soldiers in Vietnam did not expect an American victory, so why try for one?

After Vietnam the cleavages of class became increasingly difficult to reduce, and cynicism and distrust of authority became commonplace. The federal government was blamed—and bore responsibility—for such unfairness. Asked "how much of the time do you think you can trust the government in Washington to do what is right?," a whopping 76 percent of the citizenry answered "always" or "most of the time" in 1964. Four years later that figure had dropped to 61 percent, to barely more than half four years after that, and to a mere 19 percent by 1994. The proportion that told pollsters "some of the time" or "never" rose accordingly, from 22 percent in 1964 to 80 percent by 1994, an utter reversal that took the nation's leaders only a generation to accomplish. Three out of four Americans concur that "our present system of government is democratic in name only."[54]

Such a tailspin indicates an authentic crisis of confidence, to which *Catch-22* may not have directly contributed. But even before polling recorded the erosion of faith in political institutions if not the military, the novel was available to certify such distrust and to encourage a sense of some conspiracy afoot that is furtive, threatening, malevolent.[55] The belief that rationality might govern human affairs became counterintuitive.

Heller's book anticipated the notion that would seep ever more deeply into the national consciousness, the sense that the game was fixed. A glum awareness spread that the system had become an affront to the ideal of opportunity or to the promise of good fortune. "Is this a game of chance?," the rookie poker player asks in *My Little Chickadee* (1940). To which W. C. Fields slyly replies: "Not the way I play it." (The French edition of Heller's novel is entitled *L'attrape-nigaud,* or "the con game.") *Catch-22* expounded a system that is rigged; and *Something Happened* reinforced the deepening dread, the paranoia that would make its author the kid brother whom Kafka never had. The opening sentences of Heller's second novel (later put at the beginning of the second section) came to him as follows: "In the office in which I work, there are five people of whom I am afraid. Each of these five people is afraid of four people."[56] *Catch-22,* one critic noted, had "conveyed a deeply distrustful sense of modern life" through its catch-phrase of what an old woman in Rome tells Yossarian: "Catch-22 says they have a right to do anything we can't stop them from doing." Catch-22 does not even oblige those who exercise power to identify the law which they are enforcing; and this version is more synoptic than the right to be grounded which Yossarian tries in vain to assert, an effort that inspires Dr. Stubbs to declare: "That crazy bastard may be the only sane one left."[57]

Proof that the spirit animating the novel was embedded in the public culture and not limited to literary history occurred on the twenty-fifth anniversary of the publication of *Catch-22.* The most sane way for the Air Force Academy to have marked the occasion would have been to invite the author to Colorado Springs. But wouldn't such an overture, from such an institution, have been insane? Oh well. For the fall of 1986, Heller accepted the invitation, and announced that "*Catch-22* is no more anti-war or anti-military than other novels. What it's critical of is dishonesty, personal corruption, ambition—what any decent person would be critical of." The academy concurred, having assigned the novel for the past quarter of a century to brace cadets for their future in the Air Force. Heller told them that he "had the best time of my life during the war," and the hefty royalties from his first novel enabled him to quip: "I'm still making a good living from World War II."[58]

So admired a novel, so merry a mood outraged the editor of *Commentary:* nine hundred future officers were being subjected to a relentless assault on patriotic ideals and respect for authority. A book depicting "this country, and especially its armed forces . . . as being run by madmen and morons," a novel that "justified draft evasion and even desertion as morally superior to military service," Norman Podhoretz fumed, endangered military discipline. Yossarian flies seventy combat missions and earns his medal, but what mattered to Podhoretz is that he is insubordinate. What kind of readers were the cheering cadets, he wondered, if they were "incapable of recognizing a savage attack on everything they are supposed to stand for, even when it hits them in the face. If, on the other hand, the cadets were cheering Heller because they understand what he was saying in *Catch-22,* then we are producing a generation of pacifist officers who think that love of country is a naive delusion, that the military is both evil and insane, and that nothing on earth is worth dying for. God help us in either case."[59]

This plea for divine aid inspired a rebuttal from Jack M. Shuttleworth, who chaired the English Department at the academy and noted its wish to recognize achievement in the arts, of which *Catch-22* was one example. He reiterated the aim of pedagogy, which is to instill critical thinking and to expose students to ideas and values which they may not share (such as the works of Marx and Lenin, which the academy also assigned). Professor Shuttleworth also tweaked Podhoretz by quoting from his 1962 review of the novel, which the editor of *Commentary* hailed for exhibiting such "gusto and exuberance," and for "the joyful energy that explodes all over the pages of this book." *Catch-22* was simply "one of the bravest and most successful attempts we have yet had to describe and make credible the incredible reality of American life in the middle of the twentieth century." Thus it was not unreasonable to let the cadets understand somewhat better the country that they were pledging to defend. In the *Washington Post* an ad hominem letter also reminded Podhoretz of his own decision to attend college and graduate school rather than enlist in the Korean War, which cost the lives of about 54,000 United States servicemen. His autobiography admits that his 1953 draft notice was depressing, because "going into the army out of a condition of budding literary fame and growing self-confidence . . . made the experience as nearly unendurable as anything I have ever undergone." So much for the parade-ground patriotism (my kinsman, Major Major) that Podhoretz later adopted.[60]

Oddly enough the neoconservative editor did not repudiate his 1962 opinion that *Catch-22* is no pacifist novel. Even Yossarian is aware of how evil the foe is, though genocide is unmentioned. (By 1961 no specific name had been attached to the extermination of six million "non-Aryans." Not even Raul Hilberg's *The Destruction of the European Jews,* published that year, uses the term "Holocaust.") But if Heller did not elaborate on the horror perpetrated by the Third Reich, it was not because he sought to evade the implications of the pacifist position. The author never claimed to be a pacifist, and any cadets interpreting *Catch-22* as an expression of hostility to all wars would be reading it poorly (as did, it should be noted, even Alfred Kazin). Heller agreed with Podhoretz and nearly all other Americans that military resistance to Nazism had been necessary. But Heller could not formulate a rationale for intervention in the Vietnam War. Nor, while it lasted, could Podhoretz, who favored an immediate withdrawal from Vietnam only two years after Zinn's book appeared and who in 1971 preferred "an American defeat" to continuation of the struggle.[61]

But by 1986 the military needed to be spared serious criticism and satire, even as political standards could be invoked to assess fiction, as though echoing the warning of the reactionary Waugh a quarter of a century earlier that Heller's "exposure of corruption, cowardice and incivility of American officers will outrage all friends of your country (such as myself) and greatly comfort your enemies."[62] "An anti-American book," huffed Nabokov himself.[63] But one index of the caesura that the 1960s signalled was the sort of criticism Heller got from the left, "for my not being a pacifist and for Yossarian's failures to condemn that war. But these readers wanted something far beyond anything I was willing to say or feel about World War II—that any alternative is preferable to war. That's not my attitude, and it's not expressed anywhere in the book."[64] He thus honored Hawthorne's injunction to "keep the imagination sane."

That imagination proved uncanny in anticipating the lunacy and mendacity that citizens have gotten used to, as public accountability has been replaced by "plausible deniability," as feats of duplicity appear, if anything, to have gotten worse.[65] To have been wrong about the catastrophic intervention in Vietnam (or to have been silent on the subject) improved a statesman's chances for promotion in guiding American foreign policy; the erratic course of the experiment in representative government enabled Heller's novel to become an instrument of comprehension, a means of demystification. The double bind of mutually incompatible imperatives that he made notorious became a logic that the counsel for an American president proposed in the case fetchingly called *United States v. Nixon.* In July 1974 James St. Clair told the Supreme Court, as Heller himself summarized it, "that you can impeach a President only if you have evidence that he committed a crime, but you can't collect criminal evidence against a President." The claim of executive privilege was being used to thwart the efforts of the special prosecutor to gather evidence of a criminal conspiracy while Nixon's counsel was telling the Court: "A criminal conspiracy is criminal only after it's proven to be criminal." If Webster's 1970 edition defines "catch-22" as "a paradox in law, regulation, or practice that makes one a victim of its provisions no matter what one does," the White House was trying to make legal watchdogs into such victims. But as though playing Yossarian, Justice Thurgood Marshall told St. Clair, "You lose me some place along there," and joined his colleagues in rejecting such a claim of executive privilege.[66]

Just as *Catch-22* superseded the traditional form of the war novel, history redefined "realism" and turned the inconceivable into fact. The incident that the author identified as the least plausible is Minderbinder's bombardment of his own squadron, an atrocity that he gets away with: "This is the one thing that could not happen—literally. I don't think that in time of war a man could get up and actually drop bombs deliberately on his own people and then escape without punishment, even in our society." The incident defies credibility.[67] Even in Kubrick's *Paths of Glory* (1957), with the disclosure of General Mireau's order to fire on his own troops to evict them from the trenches, the disgrace is so awful that he presumably commits suicide.

During the Reagan era the founder of M & M Enterprises sprang to life as Oliver L. North, the profiteering personification of Iranscam. Milo's syndicate vows to bomb any place in Europe for cost plus 6 percent. Colonel North made operational the "neat" idea of peddling weaponry to the terrorist regime of Iran (whose Ayatollah regarded the United States as the Great Satan), to secure the funding for the contras that Congress had expressly outlawed. North's weapons were sold "to the sponsors of those who murdered 241 of his fellow United States Marines at Beirut Airport. In nations less fastidious than ours," John Gregory Dunne observed, "selling guns to the people responsible for the massacre of 241 of your country's soldiery might be construed as treason." North lied under oath to the Congress about supplying weapons to a sworn enemy—and got away with it. Hailed by President Reagan as "a national hero" (but dubbed by *Esquire* magazine a "full medal jackass"),[68] North nearly got elected to the Senate.

Compare his fate to that of Milo, who has sold ball bearings and oil to the Third Reich. He "had been caught red-handed in the act of plundering his countrymen, and as a result, his stock had never been higher." The zaniest scenario in *Catch-22* is therefore more interesting than the coincidence of a Lieutenant—later a General—named Scheisskopf metamorphosing into a General Schwartzkopf, or the way Britain's John Major reminded a commentator of Major Major, "who so lacked distinction that people who met him were impressed by how unimpressive he was," or the report after the Grenada invasion in 1983, for which more medals were awarded than the number of servicemen who had participated in Operation Urgent Fury.[69]

So "realistic" did a phantasmagoric work of fiction prove to be that news reports almost exceeded satire. Thus the Navy awarded combat ribbons to the crew of the *Vincennes,* the guided-missile cruiser that erroneously shot down an Iranian civilian airplane in 1988. A Purple Heart was also bestowed on a paratrooper who came down with heat stroke during Operation Just Cause (the 1989 invasion of Panama).[70] James Schlesinger, who served as secretary of defense under Gerald Ford, inferred from the research of behavioral scientists that "the psychological impact of a nuclear attack would result in some initial loss of confidence in government but that positive, adaptive behavior would prevail over antisocial behavior and that the survivors would support re-establishment of normal cooperative relationships at all levels of community life."

This cockamamie testimony before a Senate subcommittee was made public on January 10, 1975. Thomas K. Jones, who served President Reagan as deputy under secretary of defense, concurred and insisted that full recovery could be achieved in perhaps only two years after all-out atomic warfare with the Soviet Union. Jones believed so strongly in family fallout shelters—underground holes with three feet of dirt placed on top—that he added: "If there are enough shovels to go around, everybody's going to make it." Forget about fears of radiation: "It's the dirt that does it."[71] Thus public life has been expanding—or contorting—to meet the dimensions of Heller's antic and macabre imagination.

That's some case for a masterpiece. *Catch-22* looks rather thin were it reduced to plot summarization: it merely recounts the conflict between a bombardier and a superior officer over how many missions should be flown. Rearranged in chronological order, *Catch-22* seems rather uneventful, too: three missions to Avignon, to Bologna, and to Ferrara have all occurred before the time of the first chapter; Snowden is already dead; and M & M Enterprises has been formed. Judged as characterizations, Heller's three dozen servicemen do not exactly bulge with the three-dimensionality often ascribed to the most vibrant fiction. Most—but not all—are caricatures, and even what the protagonist looks like is sketchy. For all its scale, this novel lacks lyrical descriptions, or precise evocations of the natural world, or metaphysical depth.[72] Yet by this book we as well as our posterity are likely to know Heller, the way we also know Cervantes and Swift and Voltaire: by one book, and only one book. The cauterizing humor and pungent politics that Heller stirred together have been enduring enough, after more than three decades, to catch the reader's attention. And that's some catch, that *Catch-22*.

Acknowledgments

The author has enjoyed the opportunity to present shorter and somewhat different versions of this essay in a number of venues, beginning with the delivery of the Rolde Lecture at the Goldfarb-Farber Library of Brandeis University. Deepest gratitude is expressed to the following institutions for their hospitality: Camden County College, the Ludwig-Maximilians University of Munich, the University of Rome (La Sapienza), and the University of Wisconsin at Madison. Much appreciation is further extended to various friends, colleagues, and hosts for their criticism and their encouragement: Donald Altschiller, Christian G. Appy, Charles Cutter, Thoman Doherty, Irving Epstein, Christie Hefner, Gordon Hutner, Stanley I. Kutler, Berndt Ostendorf, Jack Pesda, Sanford Pinsker, and Michael J. Sandel. For permission to quote from papers deposited at Brandeis University, I am indebted to Joseph Heller (1923–1999), to whose memory this chapter is dedicated.

Notes

1. Markku Lahtela to Joseph Heller, April 12, 1962, in Folder 7f, Box 2 of *Catch-22* Miscellaneous Files, in Joseph Heller Papers, Special Collections, Brandeis University; Sam Merrill, "*Playboy* Interview: Joseph Heller" (1975), reprinted in *Conversations with Joseph Heller,* ed. Adam J. Sorkin (Jackson: University Press of Mississippi, 1993), 173.
2. George Plimpton, "Joseph Heller" (1974) and Chet Flippo, "Checking in with Joseph Heller" (1981), in *Conversations,* ed. Sorkin, 115, 234.
3. Herbert Mitgang, "Britain Beat Us to It," *New York Times Book Review* (October 26, 1986), 3; Judith Ruderman, *Joseph Heller* (New York: Continuum, 1991), 19–20; André Bernard, *Now All We Need Is a Title* (New York: Norton, 1994), 15; S. J. Perelman to Abby Perelman (April 15, 1954), in *Don't Tread on Me: The Selected Letters of S. J. Perelman,* ed. Prudence Crowther (New York: Viking, 1987), 164.
4. George Mandel, "Literary Conversation with Joseph Heller" (1970); and Seth Kupferberg and Greg Lawless, "Joseph Heller: 13 Years from *Catch-22* to *Something Happened*" (1974), in Sorkin, *Conversations,* 68, 122; Sarah Lyall, "For Joseph Heller, It's Finally Catch-23," *International Herald Tribune* (February 17, 1994), 20; Peter Carlson, "The Heights of Absurdity," *Washington Post* (March 19, 1998), D8.
5. Ruderman, *Joseph Heller,* 30–48.
6. Evelyn Waugh to Nina Bourne, September 6, 1961, in Folder 8f, Box 2 of *Catch-22* Miscellaneous Files, Heller Papers.
7. Norman Mailer, *Cannibals and Christians* (New York: Dial, 1966), 117.
8. John W. Aldridge, "The Loony Horror of It All—*Catch-22* Turns 25," *New York Times Book Review* (October 26, 1986), 3, and "*Catch-22* Twenty-five Years Later," *Michigan Quarterly Review* 26 (Spring 1987), 381; Alexander Bloom and Wini Breines, eds., "Past as Prologue," in *"Takin' It to the Streets": A Sixties Reader* (New York: Oxford University Press, 1995), 13.
9. James Jones to Nina Bourne (October 9, 1961), in Folder 8j; Georges Cuibus to Joseph Heller (July 13, 1964), in Folder 7g; John Steinbeck to Joseph Heller (May 7, 1963), in Folder 8g; and Thomas Pynchon to Candida Donadio (November 2, 1961), in Folder 8e; all in Box 2 of *Catch-22* Miscellaneous Files, Heller Papers.
10. Nelson Algren to Joseph Heller (December 7[, 1961]), in Folder 8a, Box 2 of *Catch-22* Miscellaneous Files; Jeremy Larner to Heller (March 28, 1962), in Folder 17:1, Box 9 of *Catch-22* Files; Kenneth Tynan to Heller, Folder 8d, Box 2 of *Catch-22* Miscellaneous Files; Stephen E. Ambrose to Heller (January 23, 1962), in Folder 17:3, Box 9 of *Catch-22* Files, Heller Papers.
11. Aldridge, "Loony Horror," 55, and "*Catch-22* Twenty-five Years Later," 382.
12. Philip Roth, "Writing American Fiction," in *Reading Myself and Others* (New York: Farrar, Straus, Giroux, 1975), 120; Norman Podhoretz, *Doings and Undoings: The Fifties and After in American Writing* (New York: Farrar, Straus, 1964), 228–29; Aldridge, "Loony Horror," 55, and "*Catch-22* Twenty-five Years Later," 382.
13. Quoted in Susan Braudy, "A Few of the Jokes, Maybe Yes, But Not the Whole Book"

(1967), in Merrill, "*Playboy* Interview"; and in Barbara Gelb, "Catching Joseph Heller" (1979), in Sorkin, *Conversations,* 37, 148–49, 197, and in Lawrence H. Suid, *Guts and Glory: Great American War Movies* (Reading, Mass.: Addison-Wesley, 1978), 270–71; Joseph Heller, *Now and Then: From Coney Island to Here* (New York: Alfred A. Knopf, 1998), 67, 169–72, and "I *Am* the Bombardier!," *New York Times Magazine* (May 7, 1995), 61.

14. Israel Shenker, "Did Heller Bomb on Broadway?" (1968), reprinted in Sorkin, *Conversations,* 54–55.
15. Paul Krassner, "An Impolite Interview with Joseph Heller" (1962), in Sorkin, *Conversations,* 8.
16. Joseph Heller in "The Books That Made Writers," *New York Times Book Review* (November 25, 1979), 80, and *Now and Then,* 88; Dale Gold, "Portrait of a Man Reading" (1969), in Sorkin, *Conversations,* 57.
17. Krassner, "Impolite Interview" and Creath Thorne, "Joseph Heller: An Interview" (1974), in Sorkin, *Conversations,* 8, 9, 132–33; Roger H. Smith, "Review," *Daedalus* 92 (Winter 1963), 155–65, in *Critical Essays on Catch-22,* ed. James Nagel (Encino, Ca.: Dickenson, 1974), 31.
18. Quoted in Suid, *Guts and Glory,* 271, and in Flippo, "Checking in with Joseph Heller," in Sorkin, *Conversations,* 232.
19. Quoted in Suid, *Guts and Glory,* 271.
20. Richard Gid Powers, *Secrecy and Power: The Life of J. Edgar Hoover* (New York: Free Press, 1987), 338–39; Thomas Kundel, *Genius in Disguise: Harold Ross of The New Yorker* (New York: Random House, 1995), 404–5.
21. Daniel Bell, "Introduction" to *The End of Ideology: On the Exhaustion of Political Ideas in the Fifties* (Glencoe, Il.: Free Press, 1960), 16.
22. Plimpton, "Joseph Heller," in Sorkin, *Conversations,* 110.
23. Quoted in Bonnie Angelo and Jordan Bonfante, "Thanks for the Memory," *Time* 135 (June 11, 1990), 10; Norman Kagan, *The Cinema of Stanley Kubrick* (New York: Grove Press, 1972), 111.
24. Charlie Reilly, "An Interview with Joseph Heller" (1979), in Plimpton, "Joseph Heller," and in Richard B. Sale, "An Interview in New York with Joseph Heller" (1972), in Sorkin, *Conversations,* 87, 108, 110, 221; Art Buchwald to Max Schuster (September 1961) in Folder 8b; Sanford Pinsker, *Understanding Joseph Heller* (Columbia: University of South Carolina Press, 1991), 24–28.
25. Robert Brustein, "The Logic of Survival in a Lunatic World," *New Republic* 145 (November 13, 1961), 11–13, reprinted in *Critical Essays on Joseph Heller,* ed. James Nagel (Boston: G. K. Hall, 1984), 30; Aldridge, "*Catch-22* Twenty-five Years Later," 380–81; Kurt Vonnegut Jr., review of *Something Happened,* in *New York Times Book Review* (October 6, 1974), 1.
26. Joseph Heller, *Picture This* (New York: G. P. Putnam, 1988), 100.
27. Murray Kempton, "A Literary Act of Faith," *New York Post Magazine* (December 26, 1961), 33.

28. Howard Zinn, *The Politics of History* (Boston: Beacon Press, 1970), 250, 258–74, and *You Can't Be Neutral on a Moving Train: A Personal History of Our Times* (Boston: Beacon Press, 1994), 87, 93–94, 97–100, 102, 110–12.
29. Richard Kostelanetz, *Master Minds: Portraits of Contemporary American Artists and Intellectuals* (New York: Macmillan, 1969), 275, 283.
30. Stanley Milgram, *Obedience to Authority: An Experimental View* (New York: Harper & Row, 1974), xii, 114.
31. Joseph Heller to Bob Houston (December 31, 1962), in Folder 17:2, Box 9 of *Catch-22* Files, Heller Papers; Daniel W. Bjork, *B. F. Skinner: A Life* (New York: Basic Books, 1993), 162.
32. Diana Trilling, *We Must March My Darlings: A Critical Decade* (New York: Harcourt Brace Jovanovich, 1977), 239; Richard Bernstein, *Dictatorship of Virtue: How the Battle over Multiculturalism Is Reshaping Our Schools, Our Country, and Our Lives* (New York: Vintage, 1995), 249.
33. Suid, *Guts and Glory,* 3, 270.
34. Suid, *Guts and Glory,* 272–73; Les Standiford, "Novels into Film: *Catch-22* as Watershed," *Southern Humanities Review* 8 (Winter 1974), reprinted in *Critical Essays on Joseph Heller,* 227–32.
35. Suid, *Guts and Glory,* 273; "Some Are More Yossarian Than Others," *Time* 95 (June 15, 1970), 74; Heller, *Catch-22,* 297; Thomas Doherty, *Projections of War: Hollywood, American Culture, and World War II* (New York: Columbia University Press, 1993), 296.
36. Brustein, "Logic of Survival," in *Critical Essays on Joseph Heller,* 31.
37. "Joseph Heller, Novelist," in *Bill Moyers: A World of Ideas* (1989), in Sorkin, *Conversations,* 287.
38. David L. Middleton, "Usually I Don't Want To Be Too Funny" (1986), in Sorkin, *Conversations,* 265; Heller, *Catch-22,* 440.
39. Quoted in Krassner, "Impolite Interview," and in David Nathan, "Catching Heller" (1991), in Sorkin, *Conversations,* 18, 294; Heller, *Catch-22,* 396; James Nagel, "Two Brief Manuscript Sketches: Heller's *Catch-22,*" *Modern Fiction Studies* 20 (Summer 1974), 221–24; David Seed, *The Fiction of Joseph Heller: Against the Grain* (New York: St. Martin's, 1989), 23, 26–27; Thomas R. Edwards, "Catch-23," *New York Review of Books* 40 (October 20, 1994), 21.
40. Quoted in Don Graham, *No Name on the Bullet: A Biography of Audie Murphy* (New York: Viking, 1989), 327.
41. Emanuel Levy, *John Wayne: Prophet of the American Way of Life* (Metuchen, N.J.: Scarecrow Press, 1988), 39–40; Richard Slotkin, *Gunfighter Nation: The Myth of the Frontier in Twentieth-Century America* (New York: Atheneum, 1992), 514–15.
42. Christopher Hanson, "The Man Who Fell to Earth," *Columbia Journalism Review* 34 (September–October, 1995), 21; "Great Triple Play in U.S. Space Race" and "A Happy End for Ham's First Flight," *Life* 50 (February 10, 1961), 17, 21.
43. Quoted in Julia Reed, "The Case of the Kissing Senator," *New York Review of Books* 43 (February 1, 1996), 24.

44. Richard H. Rovere, *The American Establishment and Other Reports, Opinions, and Speculations* (New York: Harcourt, Brace & World, 1962), 3, 18–19.
45. F. Scott Fitzgerald, *The Great Gatsby* (New York: Charles Scribner's Sons, 1925), 130.
46. Heller, *Catch-22,* 191–92; Robert Merrill, *Joseph Heller* (Boston: Twayne, 1987), 16–18.
47. David Halberstam, *The Best and the Brightest* (New York: Random House, 1972), 55.
48. Myra MacPherson, *Long Time Passing: Vietnam and the Haunted Generation* (Garden City, N.Y.: Doubleday, 1984), 141; John Gregory Dunne, *Crooning: A Collection* (New York: Simon & Schuster, 1990), 143; Christian G. Appy, *Working-Class War: American Combat Soldiers and Vietnam* (Chapel Hill: University of North Carolina Press, 1993), 28.
49. Quoted in Philip French, *Westerns* (New York: Oxford University Press, 1977), 36.
50. Dunne, *Crooning,* 141, 142, 144, 155; MacPherson, *Long Time Passing,* 123.
51. MacPherson, *Long Time Passing,* 144; Dunne, *Crooning,* 145–46; Appy, *Working-Class War,* 6, 37; Peter Levine, *Ellis Island to Ebbets Field: Sport and the American Jewish Experience* (New York: Oxford University Press, 1992), 141; John Updike, *Assorted Prose* (New York: Fawcett Crest, 1966), 110–11.
52. Dunne, *Crooning,* 149; Heller, *Catch-22,* 67; Gloria Emerson, *Winners and Losers* (New York: Random House, 1976), 216.
53. "Joseph Heller on America's 'Inhuman Callousness'" (1979), in Sorkin, *Conversations,* 211; Appy, *Working-Class War,* 185, 246–47, 317.
54. George Gallup Jr., *The Gallup Poll: Public Opinion, 1994* (Wilmington, Del.: Scholarly Resources, 1995), 32, 232.
55. Aldridge, "*Catch-22* Twenty-five Years Later," 386; Seed, *Fiction,* 59–66, 69.
56. Plimpton, "Joseph Heller," in Sorkin, *Conversations,* 106; Joseph Heller, *Something Happened* (New York: Alfred A.Knopf, 1974), 13.
57. Heller, *Catch-22,* 109, 398.
58. *Time* 128 (October 13, 1986), 73.
59. Norman Podhoretz, "Celebration with Its Own Catch-22," *Washington Post* (October 16, 1986), A21.
60. Letters to the Editor, *Washington Post* (October 25, 1986), A21; Norman Podhoretz, *Making It* (New York: Random House, 1967), 178, and *Doings and Undoings,* 229, 234.
61. Alfred Kazin, *Bright Book of Life: American Novelists and Storytellers from Hemingway to Mailer* (New York: Delta, 1974), 83; Norman Podhoretz, "A Note on Vietnamization," *Commentary* 51 (May 1971), 6, 8–9; Theodore Draper, *Present History: On Nuclear War, Détente, and Other Controversies* (New York: Random House, 1983), 358–59.
62. Waugh to Bourne (September 6, 1961), in Heller Papers.
63. Quoted in Alfred Appel Jr., *Nabokov's Dark Cinema* (New York: Oxford University Press, 1974), 66.
64. Quoted in Merrill, "*Playboy* Interview," and in Flippo, "Checking in with Joseph Heller," in Sorkin, *Conversations,* 153, 232.
65. Aldridge, "*Catch-22* Twenty-five Years Later," 386.

66. Merrill, "*Playboy* Interview," in Sorkin, *Conversations,* 171; Stanley I. Kutler, *The Wars of Watergate: The Last Crisis of Richard Nixon* (New York: Alfred A. Knopf, 1990), 509–10.
67. Quoted in Krassner, "Impolite Interview," in Sorkin, *Conversations,* 12.
68. Quoted in Theodore Draper, *A Very Thin Line: The Iran–Contra Affairs* (New York: Simon & Schuster, 1992), 548–49; "Dubious Achievements of 1987!" *Esquire* 109 (January 1988), 54.
69. Heller, *Catch-22,* 362; Nathan, "Catching Heller," in Sorkin, *Conversations,* 293; Dunne, *Crooning,* 250.
70. Philip Shenon, "What's a Medal Worth Today?," *New York Times* (May 26, 1996), 4:5.
71. Quoted in Robert Scheer, *With Enough Shovels: Reagan, Bush, and Nuclear War* (New York: Random House, 1982), 18–19, 22–23.
72. Sale, "Interview in New York," and Flippo, "Checking in with Joseph Heller," in Sorkin, *Conversations,* 80, 232; Merrill, *Joseph Heller,* 36–39; Pinsker, *Understanding,* 30–33; Seed, *Fiction,* 43–51.

CHAPTER 9

Will the Sixties Never End? Or Perhaps at Least the Thirties? Or Maybe Even the Progressive Era?

Contrarian Thoughts on Change and Continuity in American Political Culture at the Turn of the Millennium

Leo P. Ribuffo

hen Bill Clinton ran for president in 1992 as a "new kind of Democrat," President George Bush responded that there was nothing new about Clinton's outdated liberalism except an overlay of sixties decadence. Both men were campaigning in the shadow of Bush's predecessor, Ronald Reagan, who had come to office in 1981 promising, in the words of his favorite aphorism from Thomas Paine, "to begin the world over again." Although Bush claimed credit for ushering in a "new world order" with the successful prosecution of the Gulf War of 1991, many Republicans considered his record thin compared to the "second American revolution" proclaimed by Reagan. Fervent conservatives in particular distrusted Bush for failing to understand that the tax cuts of 1981 had, as Reagan declared, marked the "beginning of a new renaissance in America."[1] For his part, Clinton understood that the conservative advances culminating in the Reagan era had reduced old kinds of Democrats to caricatures for most of the electorate. Without naming names, he pointedly distanced himself from all of them except John F. Kennedy, whose assassination at age 46 at the midst of the "New Frontier" provided immunity from allegations of obsolescence.

Nothing was more characteristically American about these four presidents than their invocations of novelty and their assumption, widely shared even by conservatives who lamented the passing of "traditional" values, that newness connoted (in the language of Madison Avenue) "new *and* improved"—and perhaps benignly "revolutionary." This attitude has deep roots in the United States, a country aptly described by sociologist Seymour Martin Lipset as the "first new nation." The founders of the Republic believed that they had indeed made over part of the world and in the process had begun a "new order of the ages." Later generations followed suit by proclaiming new orders, new theologies, new souths, new lefts, new rights, and new frontiers as well as multiple revivals, reconstructions, and renaissances.[2]

The problem with these incessant proclamations of novelty is not only that they understate historical continuities, but also that they obscure what actually is new. Merely calling something new does not make it so. We must also inquire: New compared to what? Yet even historians, whose specialty should be understanding what is old, rarely bother to inquire. The tendency to accept proclamations of novelty at face value seems especially strong among the middle-aged scholars who now dominate the historical profession. Nor should we be surprised. After all, they grew up hearing that their own baby-boom generation was without peer or precedent, were subsequently influenced by "new left" intellectual currents, and now in middle age hold a vested interest in what is still called the "new" social and cultural history thirty years after its creation.

The contrarian premise of this chapter is that though American life has changed significantly during the past century, our ways of talking about politics, government, social change, and international relations show great continuity. So too does the intellectual substratum from which political leaders mine their "new" ideas. For example, during the so-called Cold War era, which stretched from late 1945 until 1989, American diplomats approached foreign policy problems created by World War II via Wilsonian premises crystallized during World War I. Starting in 1961 and for three decades thereafter, foreign and defense policy were formulated by the "junior officers of the Second World War come to responsibility" (as Elspeth Rostow, wife of John Kennedy's aide Walt Rostow, described the Kennedy foreign policy circle). The most volatile portion of the Cold War era, the "sixties," revived issues of manners and morals that had lain dormant since the twenties. And in the post–Cold War era, the new kind of Democrat in the Oval Office sought to exorcize both the sixties and the old New Deal by creating a "new progressivism" ostensibly reminiscent of the reform movement of the early 1900s.[3]

Old News: From the Nineties to the Forties

A century ago many aspects of the first new nation still looked new to western Europeans, whom Americans usually took as their reference point for measuring progress. These included mass democracy with universal white manhood suffrage, a gospel of success in which modest prosperity was promised (and often delivered) to self-disciplined families and a propensity for inflated rhetoric—what sociologist David Riesman later called "big talk." Despite occasional worries that Europeans might know something they themselves did not, most Americans took pride in their country's achievements.[4]

Yet, especially during the depression of the 1890s, fears proliferated that the United States faced acute dangers. Even mild-mannered scholars offered hyperbolic analysis. According to historian Frederick Jackson Turner, the end of the continental frontier might doom American prosperity and democracy unless "new frontiers" were found abroad. Populists and Socialists charged that upward mobility was thwarted by an emerging industrial feudalism. Physicians worried that the speed of industrialization itself induced a debilitating "American nervousness." In other quarters, threats to social and cultural order were discerned in the predominantly Catholic and Jewish "new immigration," the independent "new woman," and the liberal Protestant "new theology" that questioned evangelical orthodoxy. Even the acquisition of what Brooks Adams called a "new empire" abroad failed to eradicate such fears.[5]

The amorphous progressive movement sought—and to a large degree managed to create—a national ethos of unity, energy, and order. Whether or not the whole era is deemed progressive, the two decades before World War I affected American political culture in ways that seem permanent from the perspective of the turn of the millennium. Theodore Roosevelt and Woodrow Wilson, proponents respectively of the "New Nationalism" and the "New Freedom," presided over the establishment of a rudimentary regulatory state. Roosevelt and Wilson provided inspirational "leadership," promoted "efficiency," and denounced "special interests." These three buzz words and the popular expectations they highlight survive to the present.[6]

The so-called progressive era involved more than presidential politics and federal regulation. White southerners, including modernizing leaders of the "New South," imposed segregation and disfranchisement on African Americans. Municipal reformers curbed the political influence of new immigrants. Ratification of women's suffrage coexisted with continued hostility to new women who favored birth control and "companionate marriage." Journalists wrote exposés of current special interests and practitioners of the "new history" dissected their power in the past. The spread of nativism and white racism prompted a counterattack from cultural pluralists. More comfortable than most of his fellow citizens with the erosion of intellectual absolutes, John Dewey combined "reconstruction in philosophy" with plans for social democratic reform. Further left, bohemian rebels wanted to transform sexual mores at least as much as the means of production.[7]

Although President Wilson's belief that American ideals and self-interest usually converged in international affairs had venerable antecedents, his administration marked an important tactical turn. After vainly seeking a democratic middle way between revolutionaries and reactionaries, Wilson denied recognition to governments he judged illegitimate. The version of international order he advocated after World War I looked like progressivism writ large. The Senate refused to ratify the Versailles treaty but every president since Wilson has been a kind of Wilsonian.[8]

The domestic legacies of World War I were equally significant and more immediately obvious during the next decade. The anxieties on the home front did not subside quickly. Fear of subversion by German agents metamorphosed into fear of "un-American" activities by communists, African Americans, immigrants, bohemians, and independent women. The first Red Scare of this century disseminated a folklore of countersubversion that has been handed down to subsequent generations of "100 percent Americans." More generally, World War I turned ordinary cultural arguments into extraordinary cultural shouting matches—over Prohibition, evolution, race, and sex. There were few clear-cut victories. While enduring the ridicule of theological liberals during this century's first fundamentalist controversy, doctrinally conservative Protestants began to build their own national infrastructure. Similarly, the "New Negro" movement and Harlem renaissance reinvigorated black pride and indirectly fostered political militancy. Social scientists viewed many of the shouting matches as manifestations of a "cultural lag" among Americans slow to adjust to modernity.[9]

Major party politicians joined in the cultural shouting matches of the twenties not only because they usually adhered to one version of provincialism or another, but also because they differed so little about economic issues and needed something to say. Most agreed that Americans dwelled in a "new era" of steady prosperity and business statesmanship. Herbert Hoover, running for president as the new era incarnate, envisioned a "new day" of cooperative individualism. Although ritualistic obeisance to economic globalization lay seventy years in the future, the economies of the major nations were already sufficiently linked to pull one another down after the crash of 1929.[10]

The Great Depression and expansion of the federal government under the New Deal produced the political spectrum that remains with us despite several renovations and numerous rhetorical sleights of hand. Proponents of the welfare state usually supported the Democratic party and, following Franklin D. Roosevelt's linguistic lead, increasingly called themselves liberals while opponents of the welfare state, who were disproportionately Republican, began to formulate a coherent conservative persuasion. To the left of the New Deal, socialists and communists sought a more extensive welfare state in the present and aspired to abolish capitalism in the future. A far right, distinct from—though sometimes allied with—the conservatives, promoted a mixture of fervent nationalism and nativism. Starting in the late thirties, a congressional conservative coalition consisting of most northern Republicans and some southern Democrats thwarted major expansions of the welfare state.[11]

As citizens sorted themselves out along this political spectrum, they typically appealed to grand principles rather than to technical economic analysis, adapted cultural shouting matches in progress to fit the less prosperous and more apocalyptic context of the Great Depression, and revealed classic American hopes and fears. While conservatives accused the New Deal of coddling the undisciplined—and thus undeserving—poor, far right agitators focused on the Jews, alleged radicals, and insouciant cosmopolitans in the Roosevelt administration. Cosmopolitan "legal realists" used fifty-year old arguments against philosophical absolutism to attack conservative jurisprudence and justify prosecution of far right malcontents. Conservative intellectuals responded, as they had for fifty years, that such pragmatic relativism inevitably produced moral chaos. Virtually all shouters agreed that Americans must now become more disciplined; they disagreed about which freedoms needed restriction. While liberals and radicals urged federal regulation to discipline business, their conservative and far right adversaries urged Americans to reverse the loose living of the "roaring twenties."[12]

The Depression produced both a heightened sense that foreigners might know something that Americans did not and also, serving as an antidote, an energetic celebration of the "American way of life" past and present. The celebration was evident in popular fiction, Hollywood films, and New Deal arts projects. Meanwhile, the left wondered whether Soviets or Swedes were (as economic popularizer Stuart Chase put it) having "all the fun remaking a world." For President Herbert Hoover, Europeans illustrated what Americans should not do. Hoover viewed the New Deal as a dangerous mixture of socialism, communism, fascism, and nazism. A more celebratory conservative, Alfred M. Landon, promised to lead the nation into a bright "new frontier."[13]

World War II further invigorated the celebration of the American way of life. It resolved one major issue, unemployment, but highlighted a host of others ranging from race relations to the status of women. Republican victories in 1942 guaranteed that these would be addressed in a more conservative political context. In foreign affairs, President Roosevelt envisioned an international order led by four "sheriffs": the United States, Great Britain, China, and the Soviet Union.

Conservatives countered that the Soviets were outlaws temporarily disguised as allies. Still, there was an underlying consensus rarely acknowledged. The war against the Axis had so thoroughly discredited "isolationist" opinions that almost no one with influence doubted that the United States should lead the world.[14]

New News: From the Forties to the Turn of the Millennium

Despite its many domestic ramifications, the Cold War that followed—and to a large degree resulted from—World War II was primarily an international conflict between the United States and the Soviet Union. The conflict varied in scope and intensity during its course from 1945 to 1989. From the American perspective, the Cold War acquired an explicitly ideological cast with the proclamation of the Truman Doctrine in 1947, spread to Asia during the Chinese civil war, and became fully militarized with the outbreak of the Korean War. Crises and attempts at Soviet-American détente alternated, often in rapid succession, during the Eisenhower and Kennedy administrations. The Vietnam War temporarily disrupted the Cold War consensus. Nonetheless, the pattern of alternating crises and conciliation with some communist states persisted to the end.[15]

Throughout, however, nothing was less novel about the Cold War than the basic aspirations, strategies, and rhetorical motifs of American policymakers. Their professed goals, almost as old as the nation itself, had been codified by Wilson: a world open to democracy and free trade. Even former Vice President Henry Wallace, the premier dove of the early Cold War, looked abroad to economic "new frontiers." The assumption that the Soviet Union would ultimately collapse or mellow, the central premise of "containment" as formulated by George F. Kennan, had been standard fare among foreign-service officers since the twenties. And the description of the Soviet Union as a "totalitarian" state comparable to Nazi Germany was an intellectual legacy from the thirties.[16]

Although "totalitarianism" was an epithet rather than an analytical concept for most policymakers, they did take seriously an analogy drawn from the thirties: concessions at the Munich conference had only whetted Adolf Hitler's appetite for further conquest, and any such "appeasement" must be shunned in the contest with communism. President Dwight Eisenhower's formulation of the domino theory in 1954 was essentially a homey rendition of the Munich analogy. The Eisenhower administration claimed credit for a "new look" in defense policy. Relying on air power and nuclear bombs, the United States threatened to retaliate against communist advances anywhere in the world with an attack on the Soviet Union itself.[17]

The bombs were new but the basic premises and goals were not. American military leaders had always dreamed of technological fixes to strategic problems. If threats of massive retaliation actually deterred war, then the new look would serve the venerable goals of minimizing both American casualties and defense expenditures. The use of covert operations rather than overt military action to overthrow the governments of Iran in 1953 and Guatemala in 1954 served the same ends. In the Guatemala case, intervention in Central America was already an old look by the time Wilson left office.[18]

As a relatively coherent period within the so-called Cold War era, what is usually called the "fifties" lasted from 1947–1948, when the anticommunist consensus congealed, to 1965–1966, when opposition to the Vietnam War prompted the reopening of dormant questions about the American way of life. The accompanying second Red Scare essentially reorchestrated themes from the first: immigrants, bohemians, feminists, and militant African Americans remained prime candidates for the label "un-American." Meanwhile, liberals in public office continued their efforts to enlarge significantly the welfare state while their conservative counterparts persisted in trying to halt its growth. Neither side fully succeeded during the Truman and Eisenhower administrations.[19]

The most important political development of the fifties was the narrowing of the spectrum of reputability. Clearing their left flank, "vital center" liberals quickly repudiated communists, Popular Front liberals, and "anti-anti-communists." Clearing their right flank, proponents of a self-consciously "new conservatism" less rapidly repudiated proponents of overt religious and ethnic bigotry as well as so-called isolationists. In both cases, allusions to European ideas made old assumptions sound au courant. Centrist liberals felt more confident in labeling their conservative opponents "paranoid" because erudite refugees from the Frankfurt School diagnosed a right-wing "authoritarian personality"; for conservatives, Herbert Hoover's intuition that the welfare state slid into tyranny sounded more impressive when couched in the language of the Austrian school of economics.[20]

In retrospect, what C. Wright Mills called the "new American celebration" of the fifties looks like a continuation in a more prosperous environment of the celebration begun during the Great Depression. While cultural shouting matches were calmed or suppressed, old cultural anxieties persisted and cultural grumbling was commonplace. Prominent grumblers charged that the sources of their anxiety—juvenile delinquents, corrupt labor leaders, or relativist progressive educators—weakened the United States during the Cold War. Cold War rhetorical conventions aside, the terms of debate had changed little since before World War II. Perhaps no significant cultural development during the fifties was less novel than the fifth Great Awakening of religion. As in the previous four, clerical celebrities emerged, guides to right living proliferated, and new religions or surrogate religions were created.[21]

We must not exaggerate the impact of the Cold War on individual psyches. Although most Americans hated godless communism in the abstract and sometimes searched out subversives in local red scares, everyday life persisted with mundane tenacity. Officials seeking votes or appropriations intermittently needed to "scare the hell out of the country" (as Senator Arthur Vandenberg urged President Harry Truman as early as 1947). Yet fear was neither incessant nor unalloyed with curiosity when, for example, Soviet leader Nikita Khrushchev toured the United States in 1959. Despite Democratic efforts to frame Sputnik as the symbol of a (spurious) "missile gap," this Soviet satellite elicited wonderment as well as anxiety. The documentary filmmakers who retrospectively cherish images of suburban school children ducking under their desks during air raid drills should also show them rushing Good Humor trucks to buy ice-cream sandwiches.[22]

With Alf Landon in obscurity and Henry Wallace in disgrace, hardly anyone noticed that John F. Kennedy in 1960 was the third major political figure in this century proposing to usher in a new frontier. The compulsory vigor and informality of this New Frontier did highlight a changing political style. Even so, Kennedy's domestic agenda would have looked familiar to Landon and Wallace's old adversaries, FDR and Harry Truman. So would his defeats by the congressional conservative coalition. Kennedy's major legislative victory, the Trade Expansion Act of 1962, updated New Deal tactics for pursuing foreign markets. Similarly the "new economics" that Kennedy endorsed incorporated into public policy Keynesian ideas developed before World War II.[23]

What is usually called the "sixties" stretched from roughly 1965–1966 to 1973–1974. Certainly this segment of the Cold War era changed the "American way of life" in important ways. The supersession of that patriotic and mildly coercive phrase by the looser "lifestyle" suggests the direction of change. Yet there were important continuities too. These have been obscured because the slim historiography of the sixties exaggerates the importance of twenty-year olds.[24]

For most Americans in middle age or older, the politics of the sixties was an extension of the politics of the thirties. Even as he envisioned a "new age" of prosperity and racial equality, President Lyndon Johnson viewed the Great Society as the natural successor to the New Deal. The continuities are clear in the expansion of social security retirement benefits and creation of Medicare (a compromise liberals accepted because the congressional conservative coalition still blocked passage of national health insurance). Less obviously, the National Endowments for the Arts and Humanities were descended from New Deal cultural programs, and the emphasis on "maximum feasible participation" by the poor in the War on Poverty had precedent in New Deal "grassroots democracy." The civil rights legislation finally passed during the "second Reconstruction" had been recommended by a presidential commission in 1947; some of the militant tactics used to achieve victory had been pioneered by the Congress of Racial Equality at roughly the same time. The immigration reform enacted in 1965 similarly abolished the nationality quota system in effect since the twenties.[25]

The battles of the thirties had not ended for the right. Senator Barry Goldwater, running for president as the "new conservatism" incarnate, urged abolition of social security. As late as 1970, William F. Buckley Jr. wrote that FDR had only seemed to fashion "American ideals," but all the time he was using "alien clay."[26]

President Lyndon Johnson wanted to "thaw" the Cold War with the Soviet Union, but rejected "appeasement" of Vietnamese communism. In his own earthy version of the Munich analogy, Johnson warned, "If you let a bully come into your front yard one day, the next day he'll be up on your porch and the day after that he'll rape your wife in your own bed." Nor was this the only lesson he inferred from the thirties. Like his mentor Franklin Roosevelt, LBJ expected Americans to rally around their president during international crises and questioned the patriotism of those who did not. His attempt to suppress the antiwar movement differed from FDR's assault on the so-called isolationists only in its lack of success.[27]

Antiwar sentiment, increasing black militancy, and the re-creation of feminism as an influential movement temporarily broadened the spectrum of political respectability beyond the norms of the fifties. Champions of the "new politics," including Senator George McGovern, the Democratic presidential nominee in 1972, pursued an American dream that predated the Progressive era: creation of a left-liberal political movement that remained in contact with prevailing American ways of life. Sometimes by coincidence and almost always without acknowledgment, McGovern and other mainstream opponents of the Vietnam War echoed pre–World War II isolationists in questioning both military intervention abroad and presidential autonomy in foreign affairs.[28]

The "new left" was less concerned with prevailing American realities and was even less successful. Nor was insularity the only feature of this radical movement that looked familiar. Big talk about revolution once again found an audience, along with a renewed sense that foreigners were having all the fun of remaking the world. Still, the new left was in some respects new—or at least more reminiscent of the bohemian left of the 1910s than of the socialists and communists of the Great Depression. Although radicals abroad were again envied for remaking the world, the celebrated foreigners were now Chinese, Vietnamese, Cubans, and sub-Sahara Africans instead of the Soviets. Furthermore, the new left felt less affinity for organized labor than for the hippies, yippies, and flower children grandiosely described as a "counterculture."[29]

Issues of identity and "lifestyle" which for decades had been stigmatized as irrelevant, neurotic, or retrogressive manifestations of a cultural lag (reconfigured as "status anxiety" in the post–World War II social theory) again became central to politics in the sixties. Although the phrase "the personal is the political" is usually associated with the revived women's movement, the premise was shared by advocates of black power and gay rights as well as by evangelical Christians fighting secularization of the public sphere and working class Democrats defecting to George Wallace. Once authority figures and then authority itself were subjected to close scrutiny, primal questions about race, class, sex, patriotism, and faith began again to divide the nation. The surprise is not that these issues reemerged in the sixties but that they had been successfully pent up for so long.

Although scholars who focus on twenty-year olds typically paint the sixties as politically and culturally radical, it was instead an era of polarization. Much of the cultural shouting match—especially the shouting across generational lines—concerned the virtues or vices of newness. The signal epigram of the Berkeley campus rebellion of 1964, "Don't trust anyone over thirty," warned that no one molded by the Great Depression or World War II could understand the new world of the sixties. In *The Greening of America* (1970) counterculture theorist Charles Reich relegated most living Americans to the dustbin of "Consciousness II," as he labeled the dominant trends since the industrial revolution. On the other side, no campus confrontation was complete without at least one professor comparing student protesters to the Hitler Youth who had undermined the Weimar Republic.[30]

Although the Weimar analogy remained obscure or unknown to most Americans, those who had reached at least middle age in the sixties did bear what Caroline Bird called the "invisible scar" of the Great Depression. The identity politics of the thirties had usually involved keeping up appearances. Success might have to be deferred, but cleanliness and neatness still counted. Hence scarred survivors of the thirties rarely felt sympathy for baby boomers who not only seemed to reject success, but also tried to look like failures. Governor Ronald Reagan captured the annoyance felt by his age cohort when he described a male hippy as someone who dressed like Tarzan, looked like Jane, and smelled like Cheetah.[31]

The polarization of the sixties propelled Richard Nixon into the White House. The "new Nixon" of 1968, the second new Nixon to emerge within a decade, framed himself as a moderate between Hubert Humphrey and George Wallace. His presidency combined sordid demagogy with practical accommodation to the liberal welfare state. The accommodationist Nixon was the oldest Nixon of them all, the product of a family in which the "old progressive" Senator Robert LaFollette was admired and FDR received an occasional vote. Thus, while yearning for a "New American Revolution" limiting the federal government, this Nixon let social spending grow, expanded federal regulation, notably in the area of conservation, and experimented with a "new economic policy" of Keynesianism and wage-and-price controls. At the same time, the demagogic Nixon, a product of the second Red Scare, attempted to build an "emerging Republican majority" by appealing to race prejudice and resentment against both the "undeserving poor" and the young new left.[32]

Looking abroad, Nixon said that he wanted to combine the "idealism" of Woodrow Wilson and the "realism" of Theodore Roosevelt. Although this familiar dichotomy between Wilson and TR was dubious history, it did reveal Nixon's sense of himself as part of a diplomatic tradition. Probably the clearest parallel to TR and Wilson was Nixon's complicity in the overthrow of a Latin American government, in this case the Marxist government of Chile. In his dealings with the major communist powers, Nixon resembled instead a Republican version of Franklin Roosevelt. If he did not quite join FDR in acknowledging China as one of the world's sheriffs, at least he courted that country as a prospective member of an anti-Soviet posse. Again revealing the limits of grassroots anticommunism, most Americans willingly replaced stereotypes of China as a red and yellow peril with stereotypes of China as a potential market or exotic tourist attraction. Nixon courted the Soviets too, and he was no more candid in public about continuing Soviet-American rivalry during détente than FDR had been about continuing Soviet-American rivalry during the Second World War.

With some warrant, Nixon insisted that the Watergate scandal was no worse than similar abuses of power by Democratic presidents since FDR; they, too, had tapped, bugged, and harassed their opponents, especially in wartime. The roots went even deeper. Watergate stands out as the third Red Scare of the century. Fittingly, FBI director J. Edgar Hoover compiled his first official list of alleged subversives during the Wilson administration and his last during the Nixon administration.[33]

As a social and cultural period, the sixties component of the Cold War era ended in 1973–1974, when Nixon resigned; most American troops left Vietnam, and, above all, the post–World War II economic boom ended. Big talk of social and cultural transformation was superseded by big talk about "limits" and diminished expectations. The central economic problem—"stagflation," a combination of high unemployment and high inflation—differed from the deflationary slumps of the 1890s or 1930s. Once again, however, the response included a discourse of national decline and calls for increased discipline. Religious conservatives, especially those who sought to build a "new Christian right," anticipated God's retribution against a nation corrupted by secularization, casual sex, homosexuality, and abortion. According to dour economists, the United States needed a higher rate of savings, less consumption, enhanced productivity, and decreased job security. In this zeitgeist, the hard-working and self-disciplined Japanese led the list of foreigners now credited with refashioning the world while Americans dawdled.[34]

President Jimmy Carter tried to make the most of a difficult situation. While presenting himself as the representative of the latest (now racially tolerant) "New South," Carter also won in 1976 because he used an old Democratic tactic: portrayal of his opponent as the latest Republican Herbert Hoover. Within the limits of all such historical analogies, however, Carter shared more temperamental ground with Hoover than did President Gerald Ford. He, too, was both a technician and a humanitarian who found these two perspectives difficult to keep in focus simultaneously. In his effort to separate himself from the emerging caricature of the New Deal, Fair Deal, and Great Society, Carter reinvigorated the cardinal tenets of the older form of liberalism that had crystallized during the so-called Progressive era. Not since the heyday of Wilson and TR had a president presented "efficiency" as an essential aspect of "leadership," and no other president had ever scorned his own friendly "special interests" with such heartfelt vigor. And in his vain attempt to find a middle way between reactionaries and revolutionaries in Nicaragua and Iran, Carter replicated an old Wilsonian pattern.[35]

Yet Carter was in significant ways new and perhaps unique among modern presidents. Because as an outsider to Washington he held no vested interest in earlier foreign policy decisions, he was a routine rather than fervent Cold Warrior. More important still, no other president showed so much respect for poor, weak, or non-white nations, and none since Hoover felt less comfortable using military force. That Carter was a Wilsonian with a difference was lost on the Iranian revolutionaries who seized the United States embassy in Tehran in 1979. The difference was noticed by conservatives, who used old-fashioned language to charge the president with "appeasement" of "totalitarian" regimes.[36]

When Ronald Reagan, who became the oldest American president in 1981, cast his first vote, the major party nominees were Herbert Hoover and Franklin D. Roosevelt. Reagan's age at the time of his inauguration revealed less than his age cohort. Not only did he share the invisible scar of the Depression with most of his generation, but he also had imbibed valuable political lessons from the thirties. In ways never grasped by Carter, whose worldview developed during the more prosaic fifties, Reagan understood that the world could change suddenly rather than incrementally. In this understanding, he liked to paraphrase Tom Paine but his model was FDR.[37]

In building what came close to a new Republican majority, Reagan also applied several lessons from the fifties and sixties—especially lessons from Eisenhower's success and Goldwater's failure. First, while still resenting the "undeserving poor," most voters had developed a practical fondness for those aspects of the welfare state that helped the middle class and working class. Accordingly, except for effective efforts to undermine organized labor, the "Reagan revolution" and tax "renaissance" left most of the New Deal legacy standing. Second, while Americans overwhelmingly prayed to God, most rejected doctrinal rigor in favor or "religion in general" (a term some advocates of doctrinal rigor in the fifties had derisively applied to the fifth Great Awakening, which remained in progress when Reagan took office). Hence the "new Christian right" was relegated to a very junior partnership in the Reagan coalition. Third, while Americans were growing politically less liberal in economic policy, most had no desire to restore pre-sixties standards of decorum. Conservative politicians needed to beware of stridency and stuffiness. Commentators who called Reagan a populist were not wrong—in some sense, the term fits every president since Andrew Jackson—but they missed the distinctive feature of his political style. He was the least stilted president since Kennedy.[38]

Fourth, while most Americans went along with Wilsonian big talk in its Cold War variation, they wanted not only to stop short of the brink of nuclear war, but also to avoid enervating smaller wars like those in Korea and Vietnam. The Reagan administration fought its largest wars through proxy foreign armies in Central America. When direct intervention in the byzantine Lebanese civil war cost hundreds of American lives, Reagan promised not to cut and run, and then prudently withdrew United States forces. In retrospect, conservatives view the Reagan administration's military buildup and rhetorical offensive against communism as innovations that brought down the Soviet empire. Whether these measures helped to do so remains an open historical question, but as tactics they were virtually indistinguishable from those of Truman and Kennedy. The Reagan administration often cited these two as precedents when charged with dangerous innovations. Reagan's willingness to deal seriously with what he called the new administration of Mikhail Gorbachev suggests nothing so much as Eisenhower's opening to the Soviets after Joseph Stalin's death.

As a social and cultural period, the seventies component of the Cold War era ended during 1983–1984, when an economic upturn ended the worst recession since the thirties and Reagan solidified his coalition with a landslide victory over Walter Mondale. Meanwhile, Democratic cognoscenti incessantly urged their party to win back voters with "new ideas." The ostensibly new idea that presidential nominee Michael Dukakis advanced in 1988 was his claim to greater competence than Vice President George Bush. Benefitting from what Reagan called the "new patriotism," Republicans responded by portraying Dukakis as the sixties incarnate: a Cold War dove who also promoted big government, coddled criminals, insulted the flag, and abased himself before African American militant Jesse Jackson. In public, Bush joked that Dukakis thought a "naval exercise was something you find in Jane Fonda's workbook." Privately, Bush admitted to running, so to speak, "against George McGovern."[39]

George Bush was the last junior officer of World War II to come to responsibility as president. During the Gulf War, his comparison of Saddam Hussein and Adolf Hitler came naturally. The post–Cold War coalition Bush assembled to push back the Iraqi invasion of Kuwait did look like the fulfillment of FDR's dream of the major powers acting as "sheriffs" to enforce international order. The brief Gulf War proved more popular than Bush's vision of a "new world order," which might require continuous pursuit of international outlaws. Patrick Buchanan drew different lessons and language from the thirties when he challenged Bush's renomination in the name of a "new nationalism" and an "America First" foreign policy. Attitudes long stigmatized as isolationist won a following beyond Buchanan's constituency. Here, too, the surprise is not that these sentiments again entered national debate, but rather that they had been successfully pent up for so long.[40]

Throughout the eighties it appeared that conservatives in general and Republicans in particular would win elections indefinitely by campaigning against the radical specters of the sixties. If this tactic worked in 1988 against Michael Dukakis, a stiff relic of the fifties, then it looked even more powerful in early 1992 against Bill Clinton. Clinton presented himself as the representative of a forward-looking "new generation" as well as the son of another, even better, new South; he also proposed a "new covenant" in which the ostensibly forgotten middle class would be rewarded for hard work and moral probity. Nonetheless, Republicans responded, Clinton was a child of the sixties who had protested against the Vietnam War, avoided the draft, married a feminist, and committed adultery. Ultimately this Republican tactic proved much less durable than the Democratic habit of running against the "Hoover Depression." Probably the chief difference was that baby boomers, despite their proliferating recantations in middle age, remembered the dissident side of the sixties more fondly than their parents remembered the scars of the Depression.[41]

The domestic "new harmony" that Bush promised after his victory over Dukakis eroded during the recession of 1991–1992. Although Clinton blamed Bush for the slump, he invoked no memories of Hoover, let alone of old-style Democratic New Dealers. Instead, in high-tech idiom, he pledged to focus on the economy "like a laser beam." Bush criticized Clinton for demonstrating abroad against the Vietnam War, failing to tell the "whole truth" about his draft avoidance, and engaging in (unspecified) sinister activities while visiting the Soviet Union in 1969. Lesser Republicans assailed his wife Hillary Rodham Clinton as a foe of "family values" on the basis of articles she had written two decades earlier.[42]

These specters of the sixties might have defeated Clinton if Bush had understood how to respond, at least politically, to the recession. After all, his predecessor had ridden out a much higher rate of unemployment. But Reagan, whose father had been a local official of the Works Progress Administration, bore the invisible scar of the Depression, and thus knew that he needed to show—and personally may have felt—sympathy for the jobless. Bush, who had grown up in the small elite unaffected by the Depression, was so clueless that he vetoed extensions of unemployment benefits. Even without Clinton making the connection to Hoover, the thirties claimed a last Republican victim in 1992.

For the next four years, Clinton called himself a "new kind of Democrat" while Republicans insisted that he remained at heart a "McGovernik" (as Speaker of the House Newt Gingrich put it). Novelty lay in the eye of the beholder, and few eyes looked back very far. In fact, Clinton was the latest in a long line of prominent Democrats, including Adlai Stevenson and John Kennedy, who tried to escape the caricature of their party as the champion of the "undeserving poor," government bureaucracy, and various ethnic, racial, or working-class minorities. Attempting his own rhetorical escape, Clinton skipped over the Great Society, Fair Deal, and New Deal to draw lessons from the so-called Progressive movement. Like Theodore Roosevelt, Clinton scorned "special interests," promoted government efficiency, and tried to foster national unity. Neoliberal pundits admired his depth of historical analysis. Nonetheless, what Clinton called the "new progressivism" was essentially an intellectualization of Jimmy Carter's old intuitions.[43]

Whatever his affinities with Theodore Roosevelt, President Clinton, like his fourteen immediate predecessors, was a kind of Wilsonian in foreign policy. The United States joined other "sheriffs" in pacifying Bosnia, forcing a change of government in Haiti, and, even in the absence of the Cold War, exerting economic pressure on the Castro government in Cuba. After the Cold War, too, economics replaced military strategy as the signal foreign policy concern. Like all of his Democratic predecessors in this century, Clinton sought increased trade through lower tariffs.[44]

By 1996 the Democratic and Republican parties stood closer together on economic policy than at any time since the fifties, and perhaps since the twenties. The president and the Republican Congress not only agreed to work toward a balanced budget but also, amid bipartisan assaults on the "undeserving poor," abolished a major New Deal program, Aid to Families with Dependent Children. In 1996 Republican Senator Robert Dole became the last junior officer of World War II nominated for president. He described his campaign as a veteran's last mission. Indirectly alluding to his opponent's personal indiscretions, Dole questioned Clinton's "character." Yet he knew that it was neither dignified nor politically useful to stress the youthful misbehavior of a sitting president. Perhaps most interesting, it was the Republican rather than the Democratic candidate who fell easily into talk of the Great Depression—as a time of cleanliness and hard work. In the end, voters gave Clinton a second term to build a "bridge to the future." In the voting booth at least, the thirties finally seemed to be over and the sixties seemed to be winding down.[45]

In other areas of American life, both eras retained some hold on American minds. During a confrontation with Iraq in early 1998, Secretary of State Madeleine Albright, a refugee from Nazi-occupied Czechoslovakia, echoed George Bush in declaring, "I don't think the world has seen in a long time, except maybe Adolf Hitler, someone as evil as Saddam Hussein." Another surrogate Hitler appeared a year later, Yugoslav President Slobodan Milosevic, whose "ethnic cleansing" of Kosovo precipitated a retaliatory bombing led by the United States. While the Clinton administration and many American commentators compared Milosevic to the Hitler of genocide, Yugoslavs compared Clinton to the Hitler who had ordered the bombing of Belgrade.[46] When the latest phase in the United States–Iraqi confrontation subsided through a UN–brokered deal in 1998, hawkish critics accused President Clinton of "appeasement" reminiscent of Munich. Significantly for anyone trying to evaluate the long-term impact of the Cold War era, no one compared the agreement to Yalta. Slightly more than a decade after its end, the Cold War survives in American political culture much less as a precedent for current foreign policy, even at the level of glib analogies, than as an amorphous sense of triumph. Triumphalism is signaled above all by ritual assertions that the United States is the "only remaining superpower." By and large, these incantations betray a hope that this unique status will persist instead of a desire to replicate the crises that produced it.

Meanwhile, the "sixties" survive in the latest shouting match to define a normative American way of life—a shouting match called, with characteristic big talk, a "culture war." Some of the controversies shouted about recall the twenties (the role of the latest "new woman," the Americanization of the latest "new immigration," the place of religion in public education). Others, including abortion, have re-emerged after decades of enforced dormancy. Once again political leaders join in the cultural shouting match not only because they usually adhere to one version of provincialism or another, but also because in the absence of major differences over economics and foreign policy, they need something to say. The most tawdry shouting concerned Clinton's affair with Monica Lewinsky, which occupied the news media and three branches of the federal government for more than a year, between January 1998 and February 1999. Although Clinton was not the first president to have extramarital sex with a woman young enough to be his daughter, cultural conservatives castigated him as a distinctive representative of sixties immorality.[47]

The most enduring conflicts about the sixties occur on college campuses, radio and television talk shows, and newspaper op-ed pages. Few cultural warriors recognize or acknowledge that the general issues at hand were not even new *in* the sixties. As was the case in the Progressive era and the 1920s, politicians celebrate "efficiency" and assail "special interests" without defining either term with any precision; feminists pursue social and sexual equality while cultural conservatives lament the destruction of the family; African Americans affirm black pride and debate the merits of social integration versus separatism; a "new immigration" rapidly accepts acculturation while cosmopolitan intellectuals celebrate cultural pluralism (now renamed multiculturalism) and nativists lament a lost national identity; and scientists and fundamentalists clash over evolution in public education with as little mutual understanding as their forebears showed at the Scopes trial of 1925.

The continuities are most striking at the highest levels of erudition. Via a long detour through German philosophy and French linguistics, "postmodern" thinkers have stumbled onto the dual problem of epistemological and cultural relativism that perplexed American intellectuals from the late nineteenth century until the national celebration began in the thirties. A few of them, rediscovering John Dewey, have even rehabilitated "pragmatism" as a coherent philosophical position rather than a synonym for political wheeling and dealing. Meanwhile conservatives, who never forget (or ceased assailing) John Dewey, insist that the latest version of pragmatic relativism is especially dangerous because a "new class" of radical academics spawned in the sixties is spreading it among impressionable college students.[48]

At the turn of the millennium, both the narrow debate about politics, economics, and foreign policy and the louder shouting match to define a normative American way of life occur in three contexts. First, whether the process is celebrated as the "genius of American politics" (Daniel Boorstin's phrase from the fifties) or designated "co-optation" (a key word of the sixties), American society has lost none of its capacity to absorb and tame dissidents. Second, a journalistic style dominant since the Progressive era still tends to discover in each news cycle the greatest crisis since the sixties or the Great Depression or the Civil War when in fact the subject at hand is usually the greatest crisis since the day before yesterday. Third, although the recent per capita increase in real income falls far short of the growth rate produced during the quarter century after World War II, the stock market has risen rapidly, unemployment has dropped to the lowest level in three decades, and the federal budget has been brought into balance for the first time since the late 1960s. Viewing these developments, financial commentators discern a "new era" of prosperity without any recognition that Herbert Hoover used the same phrase in the 1920s.

So What's New?

The persistence or reappearance of old dreams, fears, phrases, and patterns of behavior should not obscure the fact that some developments are new and in a few cases significantly so. On the international scene, the United States may be the only remaining superpower, but lesser powers now possess weapons of mass destruction unavailable to them in the sixties, let alone the thirties or the early years of the twenty-first century. Moreover, though claims that American prosperity depends on foreign markets have been commonplace for more than a century, reality may have begun to catch up with rhetoric.

On the domestic scene there have also been important changes. The current cultural shouting match looks less like war than its predecessors largely because social conditions and beliefs have changed. When Americans in the Progressive era disputed whether the "new woman" represented legitimate freedom or "race suicide," few women could vote, divorces were relatively rare, and a high birth rate ensured a growing population; now women sit in Congress, roughly half of all marriages end in divorce, the birth rate stands slightly below the population replacement level, and abortion—though the most volatile "lifestyle" issue since Prohibition—has been legal since 1973. When the "new Negro" movement of the 1920s debated the merits of social integration versus separatism, most blacks were disfranchised and segregation was the law of the land; since the sixties, African Americans have achieved at least full legal equality. The old new immigration, consisting primarily of Catholics and Jews, arrived at a time when references to "Protestant Americanism" abounded; now even the "Judeo-Christian tradition," a catch phrase of toleration since the 1950s, has begun to sound exclusionary. Fundamentalists in the 1920s tried to ban the teaching of evolution from public schools; today they have fallen back on equal time for "scientific creationism."

During the past half century, self-described defenders of traditional values have changed more than liberals purporting to explore new frontiers or build bridges to the future. Some conservatives admit as much. Former Speaker of the House Newt Gingrich, the most interesting conservative ideologist to serve in high office since Herbert Hoover, considers FDR one of the three greatest Americans in this century. Erstwhile segregationists still sit in Congress but, whatever their private opinion of African Americans, in public they make do with attacks on affirmative action instead of opposing equal access to public accommodations or the ballot box. Cultural conservatives in Congress still denounce the mass media for corrupting youth, but few of them recommend restoration of the system of film censorship that survived well into the sixties—nor do many of their "traditionalist" constituents want them to do so. Liberals in retreat join President Clinton in proclaiming that the "era of big government is over." Yet the federal government is no smaller than before the "Reagan revolution." And throughout the political spectrum most Americans no more wish to abolish the welfare state than they want to ban nudity from the movies.

In short, political leaders spend most of their time engaged in symbolic battles that bypass or obscure basic economic, social, and cultural problems. Neither this phenomenon nor my complaint about it is new.

Acknowledgment

An earlier version of this article was presented at the Fourth Conference of the Americas, University of the Americas, Puebla, Mexico, in October 1999.

Notes

1. Garry Wills, *Reagan's America* (New York: Penguin, 1988), 355. Herbert Parmet, *George Bush: The Life of a Lone Star Yankee* (New York: Scribner, 1997), 467. Jack W. Germond and Jules Witcover, *Mad as Hell: Revolt at the Ballot Box, 1992* (New York: Warner, 1993). "Address Before a Joint Session of the Congress on the State of the Union," *Public Papers of the Presidents of the United States: Ronald Reagan 1985* (Washington, D.C.: Government Printing Office, 1988), 133.
2. Seymour Martin Lipset, *The First New Nation: The United States in Historical and Comparative Perspective* (New York: Basic Books, 1963).
3. Peter Wyden, *Bay of Pigs: The Untold Story* (New York: Simon & Schuster, 1979), 306. E. J. Dionne Jr., *They Only Look Dead: Why Progressives Will Dominate the Next Political Era* (New York: Simon & Schuster, 1996).

4. David Riesman, *Individualism Reconsidered and Other Essays* (New York: Free Press, 1954), 143.
5. Walter LaFeber, *The New Empire: An Interpretation of American Expansion, 1860–1898* (Ithaca, N.Y.: Cornell University Press, 1963). Donald Meyer, *The Positive Thinkers: A Study of the American Quest for Health, Wealth, and Personal Power from Mary Baker Eddy to Norman Vincent Peale* (Garden City, N.Y.: Doubleday, 1965). Sydney E. Ahlstrom, *A Religious History of the American People* (New Haven: Yale University Press, 1973), chaps. 46–47. John D'Emilio and Estelle Freedman, *Intimate Matters: A History of Sexuality in America* (New York: Harper & Row, 1988), chaps. 7–10. James S. Olson, *The Ethnic Dimension in American History* (New York: St. Martin's, 1994), chaps. 5, 8.
6. The best syntheses on the Progressive era are Arthur S. Link and Richard L. McCormick, *Progressivism* (Arlington Heights: Harlan Davidson, 1983) and Steven J. Diner, *A Very Different Age: Americans in the Progressive Era* (New York: Hill and Wang, 1998), 7. Richard Hofstadter, *The Age of Reform: From Bryan to FDR* (New York: Vintage, 1955), chaps. 4–6, and *The Progressive Historians: Turner, Beard, Parrington* (New York: Vintage, 1970). John Dewey, *Reconstruction in Philosophy* ([1920] reprinted Boston: Beacon, 1948).
8. The best distillation of Wilson's internationalist worldview remains N. Gordon Levin Jr., *Woodrow Wilson and World Politics: America's Response to War and Revolution* (New York: Oxford University Press, 1968).
9. The standard synthesis on this era is now Lynn Dumenil, *The Modern Temper: American Culture and Society in the 1920s* (New York: Hill and Wang, 1995).
10. David Burner, *The Politics of Provincialism: The Democratic Party in Transition, 1918–1938* (New York: Alfred A. Knopf, 1968). Joan Hoff-Wilson, *Herbert Hoover: Forgotten Progressive* (Boston: Little, Brown, 1975), chaps. 2–7.
11. For a good synthetic treatment of these developments, see Anthony J. Badger, *The New Deal: The Depression Years, 1933–1940* (New York: Noonday, 1989).
12. Richard Pells, *Radical Visions and American Dreams: Culture and Social Thought in the Depression Years, 1933–1940* (New York: Harper & Row, 1973). Alan Lawson, *The Failure of Independent Liberalism, 1930–1941* (New York: Putnam, 1971). Edward A. Purcell Jr., *The Crisis of Democratic Theory: Scientific Naturalism and the Problem of Value* (Lexington: University Press of Kentucky, 1973).
13. Warren I. Susman, *Culture as History: The Transformation of American Society in the Twentieth Century* (New York: Pantheon, 1984), chap. 9. Stuart Chase, *A New Deal* (New York: Macmillan, 1932), 252. Herbert Hoover, *The Challenge to Liberty* (New York: Scribner, 1934). Alfred M. Landon, *America at the Crossroads* ([1936] reprinted Port Washington, N.Y.: Kennikat, 1971).
14. John Morton Blum, *V Was for Victory: Politics and Culture during World War II* (New York: Harcourt Brace, 1976). Robert Divine, *A Second Chance: The Triumph of Internationalism in American during World War II* (New York: Atheneum, 1967).
15. In the enormous literature on the Cold War, two recent syntheses stand out as

essential: Melvyn P. Leffler, *A Preponderance of Power: National Security, the Truman Administration, and the Cold War* (Stanford: Stanford University Press, 1992), and John Lewis Gaddis, *We Now Know: Rethinking Cold War History* (New York: Oxford University Press, 1997).

16. Daniel Yergin, *Shattered Peace: The Origins of the Cold War and the National Security State* (Boston: Houghton Mifflin, 1977), chap. 1. Henry A. Wallace, *New Frontiers* (New York: Reynal and Hitchcock, 1934).
17. Stephen E. Ambrose, *Eisenhower: The President* (New York: Simon & Schuster, 1984).
18. Michael S. Sherry, *In the Shadow of War: The United States since the 1930s* (New Haven: Yale University Press, 1995), chaps. 3–4.
19. Richard M. Fried, *Nightmare in Red: The McCarthy Era in Perspective* (New York: Oxford University Press, 1990). Alonzo L. Hamby, *Beyond the New Deal: Harry S. Truman and American Liberalism* (New York: Columbia University Press, 1973).
20. Arthur M. Schlesinger Jr. *The Vital Center: The Politics of Freedom* (Boston: Houghton Mifflin, 1949). George H. Nash, *The Conservative Intellectual Movement in America since 1945* (New York: Basic Books, 1976). Richard H. Pells, *The Liberal Mind in a Conservative Age: American Intellectuals in the 1940s and 1950s* (New York: Harper & Row, 1985).
21. C. Wright Mills, *The Power Elite* (New York: Oxford University Press, 1956), 335. Mark Silk, *Spiritual Politics: Religion and America since World War II* (New York: Simon & Schuster, 1988), chaps. 1–6. William O'Neill, *American High: The Years of Confidence, 1945–1960* (New York: Free Press, 1986).
22. Richard Freeland, *The Truman Doctrine and the Origins of McCarthyism: Foreign Policy, Domestic Politics, and International Security, 1946–1948* (New York: Alfred A. Knopf, 1978). Paul A. Carter, *Another Part of the Fifties* (New York: Columbia University Press, 1983), chaps. 7, 10.
23. Herbert S. Parmet, *JFK: The Presidency of John F. Kennedy* (New York: Dial, 1983). Robert Lekachman, *The Age of Keynes* (New York: Random House, 1966).
24. For mildly revisionist accounts of the "sixties," see David Farber, *The Age of Great Dreams: America in the 1960s,* and Maurice Isserman and Michael Kazin, *America Divided: The Civil War of the 1960s* (New York: Oxford University Press, 1999).
25. Robert Dallek, *Flawed Giant: Lyndon Johnson and His Times, 1961–1973* (New York: Oxford University Press, 1999), 82. Robert A. Divine, ed., *The Johnson Years: Foreign Policy, the Great Society, and the White House* (Lawrence: University Press of Kansas, 1987).
26. Robert Alan Goldberg, *Barry Goldwater* (New Haven: Yale University Press, 1995), chaps. 4–9. William F. Buckley Jr., *American Conservative Thought in the Twentieth Century* (Indianapolis: Bobbs-Merrill, 1970), xxxix.
27. Doris Kearns, *Lyndon Johnson and the American Dream* (New York: Signet, 1976), 270. Robert D. Schulzinger, *A Time of War: The United States and Vietnam, 1971–1975* (New York: Oxford University Press, 1977). Thomas Powers, *The War at Home: Vietnam and the American People, 1964–1968* (New York: Grossman, 1973).

28. George S. McGovern, *Grassroots: The Autobiography of George McGovern* (New York: Random House, 1977), 159–252. Eugene J. McCarthy, *The Year of the People* (New York: Doubleday, 1969).
29. John P. Diggins, *The Rise and Fall of the American Left* (New York: Norton, 1992), chap. 6. Theodore Roszak, *The Making of a Counterculture: Reflections on the Technocratic Society and Its Youthful Opposition* (Garden City, N.Y.: Anchor, 1969).
30. Charles Reich, *The Greening of America: How the Youth Revolution Is Trying to Make America Livable* (New York: Random House, 1970).
31. Caroline Bird, *The Invisible Scar.*
32. Kevin Phillips, *The Emerging Republican Majority* (New York: Arlington, 1969). On the Nixon administration, see Stephen E. Ambrose, *Nixon: The Triumph of a Politician, 1962–1972* (New York: Simon & Schuster, 1989), and Joan Hoff, *Nixon Reconsidered* (New York: Basic Books, 1994).
33. Stanley I. Kutler, *The Wars of Watergate: The Last Crisis of Richard Nixon* (New York: Alfred A. Knopf, 1990) is the standard account.
34. William Martin, *With God on Our Side: The Rise of the Religious Right in America* (New York: Broadway, 1996). Lester C. Thurow, *The Zero-Sum Society* (New York: Penguin, 1981). Edward A. Purcell Jr., "Social Thought," *American Quarterly* 35 (Spring–Summer 1983), 80–100.
35. On the Carter presidency, see Peter Bourne, *Jimmy Carter: A Comprehensive Biography from Plains to Postpresidency* (New York: Scribner, 1997); Burton I. Kaufman, *The Presidency of James Earl Carter Jr.* (Lawrence: University Press of Kansas, 1993); and Erwin J. Hargrove, *Jimmy Carter as President: Leadership and the Politics of the Public Good* (Baton Rouge: Louisiana State University Press, 1988).
36. Jeanne J. Kirkpatrick, "Dictatorships and Double Standards," reprinted in Marvin E. Gettleman et al., eds., *El Salvador: Central America in the New Cold War* (New York: Grove Press, 1981), 15–39.
37. William E. Leuchtenburg, *In the Shadow of FDR: From Harry Truman to Ronald Reagan* (Ithaca, N.Y.: Cornell University Press, 1983) chap. 7.
38. On the Reagan administration, see Garry Wills, *Reagan's America*; Lou Cannon, *President Reagan: The Role of a Lifetime* (New York: Simon & Schuster, 1991); and William E. Pemberton, *Exit with Honor: The Life and Presidency of Ronald Reagan* (Armonk, N.Y.: Sharpe, 1997).
39. "Farewell Address to the Nation," *Public Papers: Reagan 1989,* 1721. Sidney Blumenthal, *Pledging Allegiance: The Last Campaign of the Cold War* (New York: HarperCollins, 1990), 286. Germond and Witcover, *Whose Broad Stripes and Bright Stars,* 403.
40. Germond and Witcover, *Mad as Hell,* chap. 8.
41. David Maraniss, *First in His Class: A Biography of Bill Clinton* (New York: Simon & Schuster, 1995) is by far the most reliable biography of Clinton.
42. Germond and Witcover, *Mad as Hell,* chaps. 24–28.
43. Colin Campbell and Bert A. Rockman, eds., *The Clinton Presidency: First Appraisals* (Chatham, N.Y.: Chatham House, 1996). Elizabeth Drew, *On the Edge: The Clinton*

Presidency (New York: Simon & Schuster, 1994). Bob Woodward, *The Agenda: Inside the Clinton White House* (New York: Simon & Schuster, 1994). Elizabeth Drew, *Showdown: The Struggle Between the Gingrich Congress and the Clinton White House* (New York: Simon & Schuster, 1997). Dan Balz and Ronald Brownstein, *Storming the Gates: Protest Politics and the Republican Revival* (Boston: Little, Brown, 1996).

44. William G. Hyland, *Clinton's World: Remaking American Foreign Policy* (Westport, Conn.: Praeger, 1999).
45. Bob Woodward, *The Choice: How Clinton Won* (New York: Touchstone, 1997).
46. "Students Receive Albright Politely," *Washington Post* (February 20, 1998), A19.
47. James Davison Hunter, *Culture Wars: The Struggle to Define America* (New York: Basic Books, 1991). William Bennett, *The Death of Outrage: Bill Clinton and the Assault on American Ideals* (New York: Free Press, 1998).
48. Richard Rorty, *Achieving Our Country: Leftist Thought in Twentieth-Century America* (Cambridge, Mass.: Harvard University Press, 1998). Irving Kristol, *Two Cheers for Capitalism* (New York: Basic Books, 1978), chap. 2. Roger Kimball, *Tenured Radicals: How Politics Has Corrupted Our Higher Education* (New York: Harper & Row, 1990). Dinesh D'Souza, *Illiberal Education: The Politics of Race and Sex on Campus* (New York: Free Press, 1991).

Contributors

CHRISTIAN G. APPY is the author of *Working-Class War: American Combat Soldiers and Vietnam* and the editor of *Culture, Politics, and the Cold War.* An independant scholar, he is currently working on an oral history of the Vietnam War.

ALAN BRINKLEY is Allan Nevins Professor of History at Columbia University and Chair of the Department of History. He is the author of *End of Reform: New Deal Liberalism in Recession and War* and *Liberalism and Its Discontents.* He is writing a biography of Henry R. Luce.

JANE SHERRON DE HART is Professor of History at University of California, Santa Barbara. She is the author of *Sex, Gender, and the Politics of the Era: A State and the Nation* and of two forthcoming books, *Litigating Equality: Ruth Bader Ginsberg, Feminist Lawyers and the Court* and *Defining America: The Politics of National Identity.*

PETER FILENE is Professor of History at the University of North Carolina at Chapel Hill. He is the author of *In the Arms of Others: A Cultural History of the Right-to-Die in America* and *Him/Her/Self:*

Gender Identities in Modern America. He is currently writing a novel about coming of age in postwar America.

JAMES GILBERT is Distinguished University Professor at the University of Maryland. He is the author of *Redeeming Culture: American Religion in an Age of Science* and *Cycle of Outrage: America's Reaction to the Juvenile Delinquent in the 1950s.* He is currently writing a book on representations of masculinity in the 1950s.

PETER J. KUZNICK is Associate Professor of History and Director of the Nuclear Studies Institute at American University. The author of *Beyond the Laboratory: Scientists as Political Activists in 1930s America,* he has just completed a screenplay about the Cold War and is currently writing a book about scientists and the war in Vietnam.

ANN MARKUSEN is Professor of Public Affairs and Planning and Director of the Urban and Regional Planning Program at the University of Minnesota's Hubert H. Humphrey Institute of Public Affairs. An economist, she is the author of *The Rise of the Gunbelt: The Military Remapping of Industrial America* and *Dismantling the Cold War Economy.* She is also editor of *Arming the Future: A Defense Industry for the Twenty-first Century.*

JOANNE MEYEROWITZ is Professor of History at the University of Indiana and editor of the *Journal of American History.* She is currently at work on a book entitled *How Sex Changed.* She is also the author of *Women Adrift: Independent Wage Earners in Chicago, 1880–1930* and editor of *Not June Cleaver: Women and Gender in Postwar America, 1945–1960.*

LEO P. RIBUFFO is Society of Cincinnati George Washington University Distinguished Professor of History. He is the author of *The Old Christian Right: The Protestant Far Right from the Great Depression to the Cold War* and *Left Center Right: Essays in American History.* He is currently writing a history of the Carter administration.

WILLIAM M. TUTTLE JR. is Professor of American History at the University of Kansas. Author of *Daddy's Gone to War* and *Race Riot: Chicago in the Red Summer of 1919,* he is currently writing a history of the G.I. Bill.

STEPHEN J. WHITFIELD is Max Richter Professor of American Studies at Brandeis University. He is the author of *The Culture of the Cold War* and *A Critical American: Dwight MacDonald and American Space*, and, most recently, *In Search of American Jewish Culture.*

Index